THE ATLAS OF APARTHEID

The Atlas of Apartheid demonstrates the highly spatial aspects of the South African government's racial policies since the National Party came to power in 1948. The study is framed by an analysis of the influence of the colonial and dominion periods and the historical and geographical legacy for a post-apartheid South Africa.

Racial separation operated at the three distinct levels of 'petty apartheid', urban segregation and 'grand apartheid'. Petty apartheid involved detailed social segregation, including separate sections of post offices and other official buildings for Whites and others. This was extended to separate recreational facilities, transport and churches. Urban segregation – most notoriously through the 'Group Areas Act' – sought to segregate place of residence and commerce. South African cities were redesigned as people of colour were moved out of the urban centres to leave zones for whites. Grand apartheid – the forced resettlement and restriction of Africans to the new homelands – confirmed the policy of excluding the majority of the population from South Africa's political process.

The *Atlas* describes the legislation and impact of apartheid and the resistance, both local and international, to its racial policies. The integration of detailed maps and text reinforces the specifically spatial nature of apartheid, the way the map of South Africa was redrawn and continues to be redrawn.

A.J. Christopher is Professor of Geography at the University of Port Elizabeth.

THE ATLAS OF APARTHEID

A. J. Christopher

London and New York

Witwatersrand University Press

First published 1994
by Routledge
11 New Fetter Lane, London EC4P 4EE

Simultaneously published in the USA and Canada
by Routledge Inc.
29 West 35th Street, New York, NY 10001

Published in the Republic of South Africa
by Witwatersrand University Press
1 Jan Smuts Avenue, Johannesburg 2001, South Africa

© 1994 A.J. Christopher

Typeset in 11pt September by
Solidus (Bristol) Limited
Printed and bound in Great Britain by
Clays Ltd, St. Ives PLC

British Library Cataloguing in Publication Data
A catalogue reference for this book is available from the British Library

Library of Congress Cataloging-in-Publication Data
Christopher, A.J.
    The atlas of apartheid / A.J. Christopher.
        p.   cm.
    "Simultaneously published in the USA and Canada by Routledge
Inc."–p. iv of CIP.
    Includes bibliographical references and index.
    ISBN 0–415–04809–5. – ISBN 0–415–10268–5
    1. Apartheid–South Africa–Maps.   2. Anti-apartheid movements
–South Africa–Maps.   3. Apartheid–South Africa.   4. Anti-apartheid
    movements–South Africa.   I. Title.   II. Title: Apartheid.
    G2566.E625C5   1994 < G&M >
    305.8′00968′022–dc20

93–24510
CIP
MAP

ISBN 0–415–04809–5 Hb
ISBN 0–415–10268–5 Pb

South African ISBN: 1–86814–237–X (Pbk)

CONTENTS

FIGURES

FIGURES

ACKNOWLEDGEMENTS

The *Atlas* developed from correspondence with Routledge in 1988 and 1989 when the lack of any comprehensive collection of maps and diagrams to illustrate the workings of apartheid was appreciated. The project progressed from there with little thought that by the time the final maps were prepared the *Atlas* would have become, in the strict sense, historical. The demise of legalized racial and ethnic segregation in the course of 1991 has been dramatic, although the inequalities left by the system are likely to persist for far longer than the majority of politicians would admit. There has therefore been that sense of urgency in the production of this book as events unfolded and the topicality of the subject was evident from the international press.

The final result owes much to the cartographic skills and patience of Wilma Stipp in the production of the maps and diagrams. The debt is most gratefully acknowledged. To Anne I express my gratitude for the assistance and encouragement she has offered during the research, writing and editing of the *Atlas*.

A.J. Christopher
Port Elizabeth, June 1992

INTRODUCTION

Apartheid has been one of the most emotive terms in the political world of the second half of the twentieth century. The Afrikaans word, apartheid, became the universally employed nomenclature for legalized and enforced racial and ethnic discrimination, notably in the fields of residential segregation, job opportunity and political rights. In its original form, derived from the parent Dutch language, the word meant 'separateness' or 'apartness'. However, in the twentieth century it assumed a political usage denoting a legally enforced policy to promote the political, social and cultural separation of racially defined communities for the exclusive benefit of one of these communities.

In 1948 the National Party came to power in South Africa and proceeded to implement the principles of apartheid within a country already deeply immersed in colonial segregationism. The consequences were far-reaching, with large-scale forced removals of population, and the redrawing of the internal political structures of the state. Race became the dominant element in determining the rights, political and legal, of members of the population. The map of South Africa and its towns and cities were redrawn on racial lines with different rights assigned to different groups of people within the different zones.

The whole process was fuelled by the determination of the politically and economically dominant White group to retain power over the country in the face of rising demands for political rights by the Black majority. Unlike the other mid-latitude British dominions, the immigrant populations never achieved demographic dominance as opposed to political and economic domination in South Africa. At its relative height in the first half of the twentieth century the White population only amounted to a fifth of the total for the country as a whole (Figure I.1). This proportion has since declined to stand at approximately 13 per cent in 1991.

Economic dominance was most clearly illustrated by the disparity in income and spending power between the various sectors of the population. In 1946 the White one-fifth of the population controlled two-thirds of personal income (Figure I.2). Only in the 1980s did the White share of income begin to fall significantly. The enforcement of racial job reservation, and hence maintenance of high White incomes, was emotionally

1

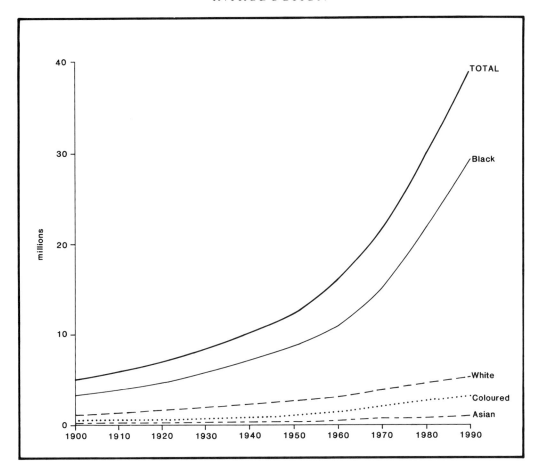

Figure I.1 Population of South Africa, 1900–90

Source: Statistics from census reports 1904–80 and projections to 1990

linked to the preservation of continued White rule. This was baldly stated by Mr B.J. Schoeman, Minister of Labour, in 1954, in the debate on the Industrial Conciliation Bill:

> this provision is against economic laws. The question however, is this: What is our first consideration? Is it to maintain the economic laws or is it to ensure the continued existence of the European race in this country.... I want to say that if we reach the stage where the Native can climb to the highest rung in our economic ladder and be appointed in a supervisory capacity over Europeans, then the other equality; namely political equality, must inevitably follow and that will mean the end of the European race.[1]

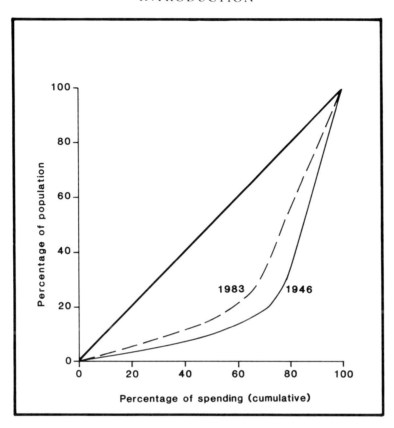

Figure I.2 Per capita spending by population groups, 1946 and 1983

Source: Statistics from H. Adam and H. Giliomee (1979) *Ethnic Power Mobilized: can South Africa change?* New Haven: Yale University Press, and South African Institute of Race Relations (1989) *Race Relations Survey 1988/89* Johannesburg: S.A.I.R.R., p. 423

The privileged political and economic position of the White population was kept in place through the ruthless enforcement of an elaborate series of laws which were constantly amended and extended to remove any loopholes which might be discovered. Many of the enactments, including the key Natives Land Act (1913), the Native Trust and Land Act (1936) and the Natives (Urban Areas) Act (1923), predated the National Party's accession to power, but were rigorously enforced and amended to serve new purposes after 1948. Other legislation was entirely new, notably the Population Registration Act (1950), the Group Areas Act (1950) and the Promotion of Bantu Self-government Act (1959), which sought to establish a more comprehensive regime. Numerous other enactments determined in detail the activities of the majority of the population, the bureaucracy was enlarged to administer the legislation, and the police force extended to supervise obedience to the law.

The apartheid legislation was initially designed to establish urban racial residential segregation. The Prime Minister, Dr D.F. Malan, was most emphatic that this was the key to apartheid, when debating the Group Areas Bill in 1950:

> The Group Areas Bill ... is a measure which is proposed by this side of the House to carry out its policy of apartheid, and in that respect it undoubtedly is the most important of all ... will mean a fresh start for South Africa, because it heralds a new period. I do not think there is any other Bill, affecting the relations between the different races, the non-Europeans and the Europeans in this country, which determines the future of South Africa and of all population groups as much as this Bill does.[2]

Subsequently attention was directed towards the exclusion of anyone not classified as White from the political process of the country. The Population Registration Act thus formed the cornerstone of the entire policy as everyone was required to be classified into a distinct racial group. In such a system there was no room for ambiguity. In 1949 the government had passed the Mixed Marriages Act, with the object, in the words of Dr T.E. Donges, Minister of the Interior: 'to check blood mixture, and as far as possible promote racial purity'.[3] Not content with a dual White–Black grouping, the government introduced a complex classification which sought to divide the indigenous population from those of mixed parentage, and other immigrant groups. Furthermore, the indigenous population itself was subdivided, according to home or inherited language, into a series of national 'units' or incipient nations.

Separation on the personal level was deemed essential for the survival of the White race. A series of laws were enacted in order to prevent contact between Whites and other people. The Minister of Justice, Mr C.R. Swart, introducing the Reservation of Separate Amenities Bill in 1953 sought to show why such measures were indispensible:

> If a European has to sit next to a non-European at school, if on the railway station they are to use the same waiting-rooms, if they are continually to travel together on the trains and sleep in the same hotels, it is evident that eventually we would have racial admixture, with the result that on the one hand one would no longer find a purely European population and on the other hand a non-European population.[4]

It might be added that amenities were to be separate but not equal:

> it was never the intention of Parliament to say ... that if you reserve something for one group, equal provision should be made in every respect for the other group. In our country we have civilized people, we have semi-civilized people and we have uncivilized people. The Government of the country gives each section facilities according to the circumstances of each.[5]

Inequality was most marked in the realm of education, where Dr H.F. Verwoerd, when Minister of Native Affairs, piloted the Bantu Education Bill through parliament in 1953:

Racial relations cannot improve if the wrong type of education is given to Natives. They cannot improve if the result of Native education is the creation of frustrated people who, as a result of the education they received, have expectations in life which circumstances in South Africa do not allow to be fulfilled immediately, when it creates people who are trained for professions not open to them.[6]

Echos of the Nazi education policies for the subjugated peoples of Eastern Europe in the early 1940s are evident. Martin Borman, Hitler's party secretary, summed up the attitude succinctly: 'Education is dangerous. It is enough if they can count up to 100.... Every educated person is a future enemy'.[7]

The ultimate apartheid experiment was the excision of the Black areas from the country, and hence the removal of the Black population from the political process. At first the National Party government was ambivalent on the issue, prompting one author to suggest that 'The Myth of the "Grand Design"' had been propagated politically but was not consistent with the record.[8] Dr D.F. Malan in 1954 envisaged the possibility of such a scheme in the distant future:

Theoretically, the objective of the policy of apartheid can be fully realised by dividing the country into two states, with the whites in the one, and all the natives in the other.... Whether in time to come we shall reach a stage where a division of this nature ... will be possible, is a matter we have to leave to the future.[9]

By 1959 the new Prime Minister, Dr H.F. Verwoerd, had injected a greater sense of urgency and direction into debate on the political future of the Black population:

I say that if it is within the power of the Bantu and if the territories in which he now lives can develop to full independence, it will develop that way. Neither he nor I will be able to stop it and none of our successors will be able to stop it.[10]

Such was the success of Dr Verwoerd's partition policy that by the 1970s several Black political leaders had accepted the concept of independence from South Africa and were concerned with how much territory they could obtain. President Lucas Mangope of Bophuthatswana argued for a more generous division of the country in order to obtain international credibility at his country's independence celebrations in 1977:

Independence and consolidation are only the two sides of the same coin. If any one side of this coin lacks integrity and credibility, the coin will be regarded as faked, and it will be rejected. It is quite self-evident, that the Achilles' heel of Bophuthatswana's credibility is the present state of consolidation, or rather non-consolidation.[11]

Apartheid in its most basic form involved the removal of anyone not considered to be White from areas deemed to be for the occupation or enjoyment of the White population. Accordingly massive forced population movements in both the urban and rural areas began in the initial years of the National Party's administration. The reorganization of South Africa's towns and cities was a long-term process extending over more than two decades. Similarly, the removal of the 'surplus' Black population from the 'White' rural areas was spread over as long a period of time. The lengthy time span allowed for an evolution of the apartheid concept from its basic segregationist form to the more complex concepts of 'separate development' and 'separate freedoms', even if the intent of maintaining White supremacy remained the same.

Apartheid was thus conceived as a spatial policy, with markedly geographical consequences. Lines were drawn on maps at various scales, and people were evicted and resettled to fit the lines. The administrators were acutely place conscious with a heightened emphasis on the distinctiveness of particular places and communities. Not only were towns redesigned into separate sectors, even with separate administrations, but an entirely new political map of the country unfolded as the policy developed. State-partition became the official aim by the 1970s, with South Africa fragmented into a series of Black nation-states and a large White-controlled rump entity. Policies were pursued to increase the political and economic viability of the proposed Black nation-states, but without weakening the position of the White state. Industrial development, transport planning and regional planning were all undertaken with the goal of creating a new, smaller, but whiter, South Africa.

Problems increased as White minority rule became more untenable in the final quarter of the twentieth century. International ostracism, economic sanctions and external military involvement on the one hand and the growing internal political and economic pressures and problems on the other, led to reappraisals in the 1980s. However, official attempts aimed at the 'modernization' of apartheid met with little success. Relative White demographic decline, particularly highlighted by the massive rural–urban migration of the Black population in the 1980s, illustrated the practical impossibility of implementing the entire apartheid policy. The end was thus far more sudden than expected, following the epoch-making opening of the new session of parliament by President F.W. de Klerk on 2 February 1990. The massive legislative and administrative burden of the apartheid structure was removed rapidly in the course of 1991 and negotiations for a new dispensation began between the interested political parties and organizations, many of which had previously been banned. The overwhelming rejection of apartheid, including the concept of state partition, by White voters in a referendum in March 1992, prompted President de Klerk to state that 'South Africa has closed the book on apartheid'.[12]

However, the physical and social heritage of over forty years of enforced separation will only be seen in the coming decades as a post-apartheid society comes into being. The legacy of the policy will be a significant impediment to the country's future development. Indeed the Reverend Alan Hendrickse, Leader of the Labour Party, subsequently visualized apartheid still ruling from the grave, as its formal dismantling was not sufficient to right the wrongs perpetrated in the era. Such fears appeared to be justified in view of the convoluted thinking behind the 1992 amendment to the Defence Act, which amongst other items stated that:

'White person' means a person who (a) in appearance obviously is a white person and who is generally not accepted as a coloured person; or (b) is generally accepted as a white person and is not in appearance obviously not a white person, but does not include any person who for the purposes of his classification under the repealed Population Registration Act freely and voluntarily admits that he is by descent a Black or a coloured person unless it is proved that the admission is not based on fact. The above definition of 'white person' must be read with the following definitions: (a) 'coloured person' means a person who is not a white person or a Black; (b) 'Black' means a person who is, or is generally accepted as, a member of any aboriginal race or tribe of Africa.[13]

The period from 1948 to 1992 thus constitutes a remarkable era in world history, when South Africa occupied a particularly unenviable position as an international pariah state in consequence of its internal policies. *The Atlas of Apartheid* seeks to demonstrate the spatial patterns of the planning and enforcement of the policy.

State policy operated at three levels. 'Grand apartheid' sought to partition the country in an effort to ensure continued White control of the remainder. 'Urban apartheid' sought to ensure that the living and business areas of the cities were segregated. 'Petty apartheid' sought to ensure minimal personal contact between peoples of differing racial classifications, through aspects ranging from separate entrances to buildings to separate schools. Each of these was concerned with the control and use of physical space.

Politicians and bureaucrats drew lines upon maps at many scales in order to separate people into neat racial compartments. The essentially spatial characteristics of apartheid lend themselves to interpretation through the presentation of maps. Consequently in an era of rapid political change an atlas becomes a vital reference work for comprehending the nature of the transition in progress.

The Atlas of Apartheid is not intended to be a chronological account of the history of South Africa between 1948 and 1992. Many and conflicting such accounts have been and no doubt will be written. It is intended here to illustrate the spatial aspects of the policy of apartheid, demonstrating the essentially geographical planning which was effected in the era. The work is therefore organized topically around broad themes in an attempt to bring together the various and diverse aspects of the South African state's apartheid project. Some of the more general tenets of apartheid will consequently be absent, where the inequalities are not spatially reflected, or where statistics are not available.

The collation of statistics on a racial basis, often excluding the Black population or referring only to the White population, makes any work on South Africa particularly problematical. Territorially, too, the term 'South Africa' has meant different extents, ranging from the broad concept of South Africa including Botswana, Lesotho and Swaziland, to the narrowly focused White apartheid state, excluding all the Black homelands, whether 'independent' or not. In addition, for most of the period since the mid-1960s, the country has been subject to international ostracism and economic sanctions, thus much statistical material has been regarded as sensitive and therefore not available to the researcher. Political turmoil since 1976 has further rendered the results of all post-1970 population censuses suspect.

If examples from Port Elizabeth appear frequently in the text, this is not because the city was the most segregated in the country or had any other distinctive feature, but because the author is most familiar with the peculiarities of the apartheid system as manifested in this city. Accordingly it is hoped that readers may obtain a greater appreciation of the intricacies of the policy through familiarity with a single city.

There will be no concluding chapter – the maps and text speak for themselves of a uniquely selfish and degrading concept.

1

BEFORE APARTHEID

The development of apartheid was firmly rooted in the colonial era. The country was first settled by European colonists under the Dutch East India Company in 1652. In 1806 the colony was occupied on a permanent basis by Great Britain. Although at the end of the Dutch period there were only 25,000 European settlers in the country, they had through the evolution of an extensive farming system occupied land up to 1,000 kilometres from the original settlement at Cape Town. Within the isolation which this system imposed, a distinct community evolved, later to become the Afrikaner nation. In the British period, South Africa was not perceived as a land of opportunity to rival the United States, Canada, Australia or New Zealand as a destination for emigrants. As a result European immigration was never on a scale whereby British immigrants out-numbered the Afrikaners. Indeed between 1815 and 1914 less than 4 per cent of emigrants from the United Kingdom went to South Africa (Figure 1.1). However, the British contribution in administrative and commercial fields was profound.

Conflict between the British authorities and sections of the Afrikaner community seeking political independence was a constant theme throughout the history of the nineteenth century. Thus on a political level the establishment of the Union of South Africa in 1910 as a British dominion was viewed as a compromise between the two White 'races', which consequently had nothing to offer the indigenous population in the form of political and economic prospects.

It would be an understatement to say that the indigenous population did not fare well under the colonial regime. The Khoisan peoples of the western half of the country, who came into contact with the colonists first, were worst affected. The population was sparse, as befitted a herding and hunting economy. Loss of lands and livestock as a result of the steady encroachment of the White colonists reduced most of its members to servitude, as tenant labourers on the newly established European owned farms. Disease, notably smallpox, also took its toll, severely decimating numbers in the first half of the eighteenth century. Thus in the western half of the country the indigenous population, reduced in numbers and deprived of land and livelihood, had shrunk to a small minority in many districts by the early twentieth century (Figure 1.2). In this respect the western half of the country followed the experience of many other

9

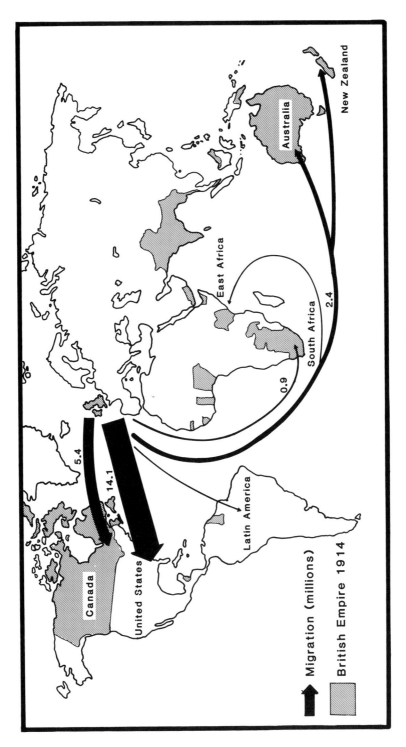

Figure 1.1 British emigration flows, 1815–1914

Source: From statistics in A.J. Christopher (1988) *The British Empire at its Zenith*, London: Croom Helm

mid-latitude colonies, notably the Americas and Australasia, where imported pathogens worked their deadly effects. However, in the eastern, better watered, sector of the country the colonists encountered the Bantu-speaking peoples, with a mixed sedentary agricultural base. They were less susceptible to European diseases and proved to be difficult to control militarily, and so dispossess of their land and livestock. Consequently, there ensued a long period of conflict from the 1770s to the 1870s, while colonial and local governments conquered the various indigenous chieftaincies and kingdoms of the subcontinent in a piecemeal fashion.

In 1910 the Union of South Africa was established through the unification of four British colonies (Figure 1.3) At that time there were some five million people within the

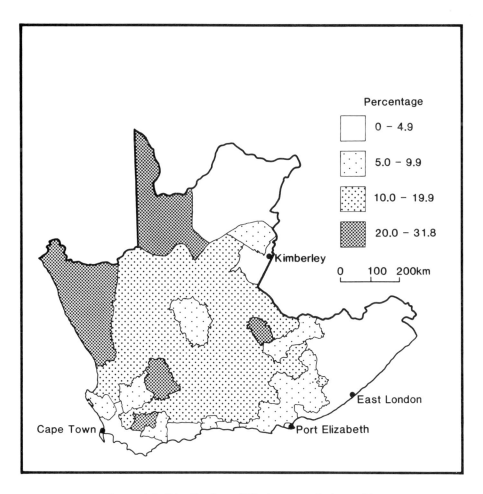

Figure 1.2 Distribution of Khoisan population, 1904

Source Computed from Cape of Good Hope (1905) *Census of the Colony 1904*, Cape Town: Government Printer

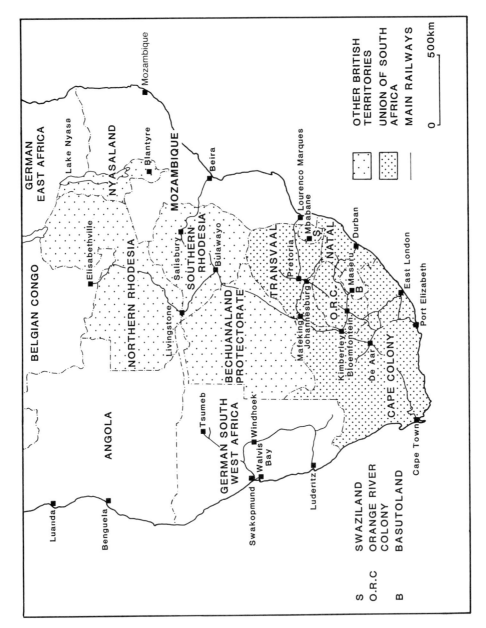

Figure 1.3 Southern Africa, 1910

Source Compiled by the author from various sources

boundaries of the new dominion, of whom only one million were of pure European origin. It was, however, this one million which controlled the government and remained determined to continue to exercise exclusive political power in the ensuing eighty years. The methods employed for doing so changed from the pragmatic segregationism of the pre-1948 period to the ideologically formed apartheid policies of the post-1948 era. Although a distinct break at this date is often implied, there is a remarkable degree of continuity from the colonial period through the Union segregationist era to the age of apartheid.

COLONIALISM

The maritime colonial powers initially bypassed South Africa in their quest for power and profit. The Portuguese, although mapping the coastline, did nothing to establish permanent bases on this section of the continent, as there were no recognizable exploitable resources and the indigenous population appeared hostile. Only when the advantages of establishing a provisioning station for ships voyaging between Europe and the East Indies were perceived, did the Dutch East India Company decide to act. In 1652 a settlement was founded at Cape Town to provide fresh fruit, vegetables and meat to the company's vessels. A garden was laid out to serve these needs. However, the company's limited resources were such that it was not possible to fulfil the expectations and recourse was made to the settlement of free individual European settlers to grow the required provisions. The supply of meat was also problematical as the indigenous population proved reluctant to barter their cattle away in the numbers required. Conflict and seizure ensued, with the company, and later individual settlers, undertaking cattle, sheep and goat raising on lands taken from the indigenous population. Attempts to define the physical boundary of the settlement, in order to limit the extent of conflict, began with the planting of a hedge of bitter almonds instigated by Jan van Riebeeck, the first Dutch commander (Figure 1.4). This, as all subsequent demarcations, proved to be ephemeral as European settlers seized the opportunities which were offered by a new land with minimal government interference and an initially limited indigenous opposition.

The colonial scale of operations thus increased rapidly as extensive systems of cultivation and stock farming replaced inherited intensive European models. A constant lack of labour and capital hampered development, as all the colony appeared to offer in abundance was land. Settlers from the Netherlands, Germany and France were placed on small (20–50 ha) farms in the seventeenth century. However, in 1717 the government halted free immigration as the colony was considered to be fully occupied, although it contained only 2,000 European inhabitants. The limited physical resources of the immediate hinterland of Cape Town restricted commercial agriculture to grain and wine enterprises, which were run on tropical plantation lines with a slave labour force imported from the Dutch colonies in Asia. The vital decision as to whether to run the colony with free or servile labour was taken in 1658, effectively determining the

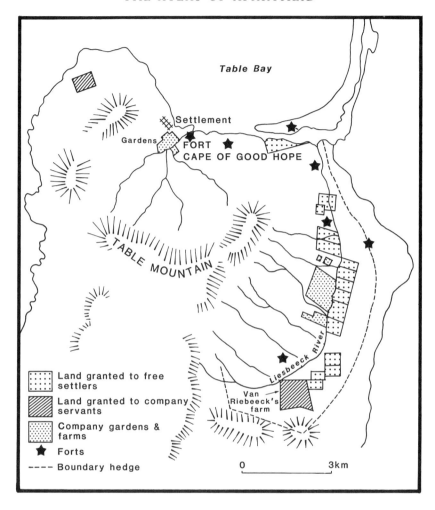

Figure 1.4 The first Dutch settlement at the Cape of Good Hope

Source After A.J. Christopher (1976) *Southern Africa*, Folkestone: Dawson

form of colonial society, and thereby influencing the social attitudes and population composition of later generations.

Parallel with these developments was the evolution of extensive livestock raising in the interior of the colony. The seasonal movement of livestock to lands in and beyond the mountains of the Western Cape, evolved by the early eighteenth century, was subsequently transformed into the permanent establishment of independent stock farms in these regions. Farms of 2,500 hectares were offered at a nominal rental to those venturing on to the frontier of European settlement. The result was the rapid appropriation of the best agricultural and grazing land across the subcontinent and the dis-

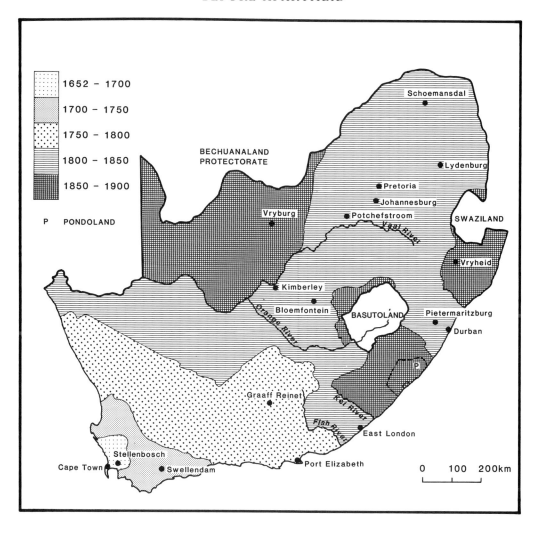

Figure 1.5 Sequence of European annexation

Source Compiled by the author from various sources

possession of its occupants (Figure 1.5). By the late eighteenth century lands were being settled over 1,000 kilometres from Cape Town, and the settlers had come into conflict with the indigenous agriculturalists, whom they were unable to displace on any significant scale with the technology available to them. Settler pastoralism differed little from that of the indigenous population whom they incorporated into the system wherever possible. Opportunities for commercialization were limited as distance precluded the marketing of more than small quantities of livestock and livestock products in exchange for luxury items, firearms and wagons. In the main, settler society

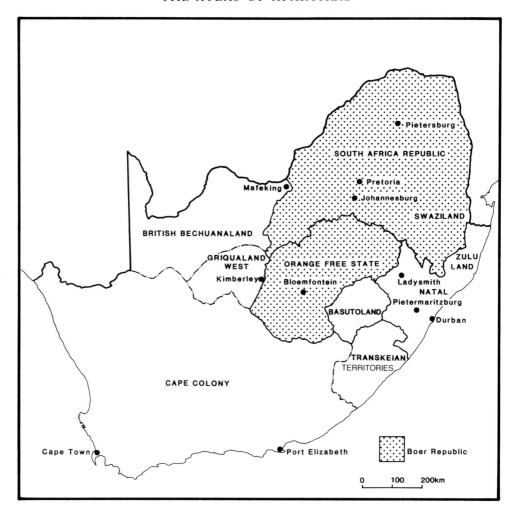

Figure 1.6 Political map of South Africa, 1899

Source Compiled by the author from various sources

was self-contained with only limited contacts with the commercial world or indeed between individual members. Isolation was the overwhelming result of the extended lines of communication to and from Cape Town and the poverty of the physical environment.

In the nineteenth century the process was continued. Afrikaner settlers pushed into the interior of South Africa across the Orange River, and entered areas initially considered to be virtually devoid of indigenous people as a result of the massive upheavals of the *mfecane*, associated with the rise of the Zulu military monarchy. In the late 1830s some 15,000 Afrikaner colonists migrated north and eastwards to occupy large tracts of the Orange Free State, Transvaal and Natal. In these regions they established their own independent republics, free of British control. Chaos in Natal led to British annexation and settlement by new immigrants, but the Orange Free State and Transvaal, after involved political manoeuvring, survived as separate entities until the end of the century, when they again came under British rule. The Boer War (1899–1902) is one of the epic conflicts in the building of an Afrikaner mythology and sense of identity. The political map of 1899 is thus still able to evoke political emotions as representing the Afrikaner homeland (Figure 1.6).

Within the republics land was appropriated in a similar fashion to that evolved in the Cape Colony. The same semi-self-sufficient economy was reproduced, although ivory possibly contributed more to local funds than agriculture as such in the first decades of settlement. Commercialization of farming was only possible with the development of wool exports in the 1830s and 1840s in the Cape Colony, and the extension of sheep farming elsewhere in the ensuing decades. However, in much of the Transvaal sheep were subject to diseases which precluded their commercial exploitation and no viable substitute was introduced until the twentieth century.

Within these new states and colonies the indigenous population remained in substantial numbers, both on the periphery, where effective settler control was only established late in the century, and in the European occupied farming area, where substantial workforces were acquired along with the land. So large were the indigenous populations incorporated within the colonial and republican state boundaries from the 1830s onwards that a new form of organizational control was introduced. Rural reserves were set aside where the indigenous population could continue to live under traditional rulers and customs but overall European supervision. In some cases the lands were granted to co-operative leaders who assisted the settlers or the governments, while in others they were demarcated to accommodate those considered to be 'surplus' to the labour requirements of the settlers.

Later, as the colonies and republics incorporated regions which were densely occupied by the indigenous population, little change was made in the land dispensation as virtual protectorate status was accorded to the indigenous social hierarchy. In all cases it needs to be emphasized that the land so set aside was regarded as belonging to the government, to be taken away if the conditions of agreement were infringed. No indigenous land rights were recognized as the basis of a colonial or republican title deed. Conquest and annexation effectively transferred sufficient rights to the Crown or the State to ensure that all further land holdings derived their rights from those authorities. However, attempts to draw a permanent line to define the boundaries between immigrant and indigene remained a constant theme of nineteenth-century politics, as they had done since the foundation of the first Dutch colony, and were to become an integral part of the apartheid policy.

THE POPULATION

The population of South Africa is highly complex both in its origin and distribution. The indigenous population, broadly divided into the Khoisan population of the western regions, and the Bantu-speaking population of the eastern regions, constitutes the numerical majority, while the immigrant peoples are most highly concentrated into areas of specific economic opportunity, notably the towns and restricted rural areas outside the defined reserves (Figure 1.7).

In any discussion of population in South Africa the racial connotation is of paramount importance, as membership of a particular racial or ethnic group has largely defined the individual's opportunities and rights. Thus throughout the colonial and Union periods racial classification has been of the utmost significance. At first the basic distinction was made between Christian and Heathen. Later this was amended to Free and Slave, then the terms 'European' and 'Coloured' were adopted to denote the same divide. Such bi-polar divisions became increasingly fragmented in the nineteenth century, when under the influence of anthropological taxonomy ever more elaborate classifications were used in the censuses. Accordingly in the last Cape Colonial census in 1904 seven major categories were employed, fragmented into over forty sub-divisions. Although simplified at Union, the census commissioners continued to employ a racial classification system, based on physical and social characteristics. The problems of such a classification may be illustrated by the grouping 'other, mixed and other' in 1904. Broadly some four groups were distinguished in the Union period, namely: White or European, Native (later Bantu or Black), Coloured or Mixed, and Asian. As evidenced in other classificatory censuses, notably in British India, the categories devised for taxonomic purposes assumed legal and political status.

The White population of European origin constituted a minority throughout South Africa by 1951 (Figure 1.8). It was derived from a range of countries, but broadly fell into two groups, the Afrikaans and English speakers, who controlled the political process. Substantial German and later Portuguese communities were also present. It should be remembered that the majority (approximately 55 per cent) of the European population was Afrikaans speaking and that after 1910, all South African prime ministers and presidents have come from that community. With the exception of restricted parts of the eastern Cape and Natal all the European-owned rural areas were dominated by the Afrikaans speaking community (Figure 1.9). In contrast the cities, because of the pattern of colonial economic development, were largely dominated by English speakers until the 1950s.

The numerically dominant, officially defined indigenous Native, African or Black population was composed only of Bantu speaking people, who had not penetrated the western Cape by the beginning of the colonial era (Figure 1.10). Owing to the long history of occupation the various communities developed separate languages, often mutually unintelligible. Three broad ethno-linguistic divisions are represented in the country: the Nguni peoples of the eastern coastal regions, the Sotho peoples of the

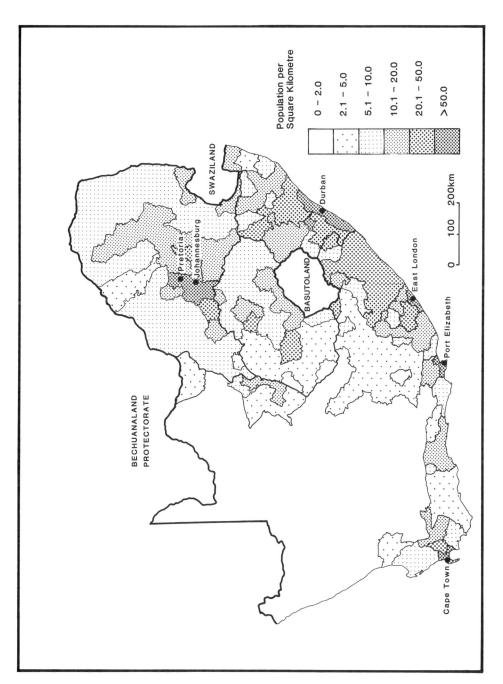

Figure 1.7 Distribution of population, 1951

Source Computed from South Africa (1955) *Population Census 1951*, Pretoria: Government Printer

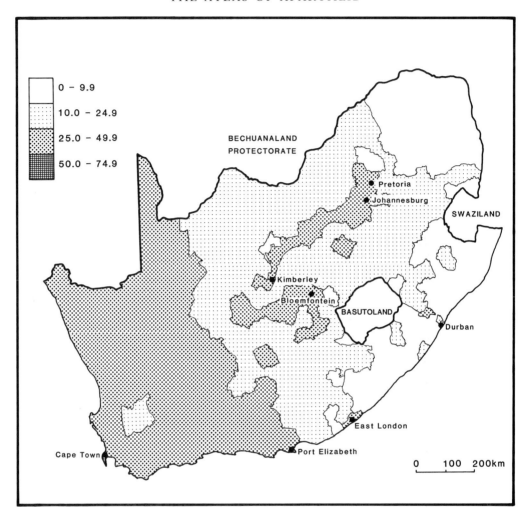

	0 - 9.9
	10.0 - 24.9
	25.0 - 49.9
	50.0 - 74.9

Figure 1.8 Percentage of population returned as White, 1951

Source Computed from South Africa (1955) *Population Census 1951*, Pretoria: Government Printer

interior, and the Vendans and Shangaans of the extreme northern and eastern Transvaal. In pre-colonial times these peoples were subject to a range of authorities from the centralized Zulu monarchy to the individual village headmen, subject to only a tenuous higher authority.

The Asian population was largely descended from Indians who initially came as indentured labourers to work on the sugar plantations of Natal in the nineteenth century. Later others immigrated from that subcontinent as traders and professionals to serve the community, which consequently reflects the religious, linguistic and cultural diversity of India. Owing to the remarkable commercial enterprise of some members of

20

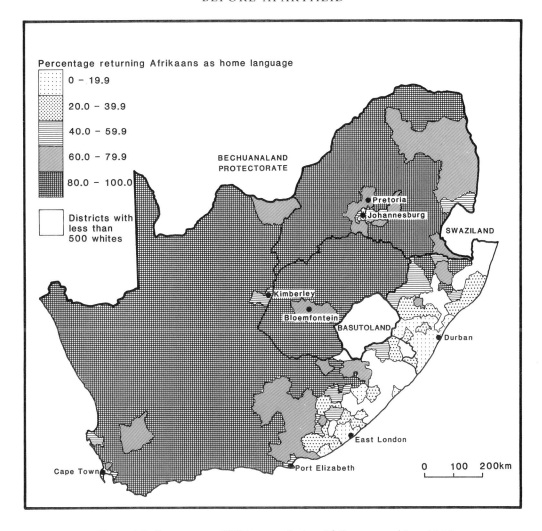

Figure 1.9 Percentage of White population Afrikaans speaking, 1946

Source Computed from South Africa (1954) *Population Census 1946*, Pretoria: Government Printer

the community, the Indians have always maintained a high profile in European politics when European commercial interests were perceived to be threatened by industrious competition. The community was predominantly urban by 1951 and so constituted a significant part of the population of the major centres of Natal and the southern Transvaal and to a lesser extent the Cape coastal cities (Figure 1.11). Smaller numbers of Chinese emigrated to South Africa between 1890 and 1949 and form a distinct sub-group largely engaged in trade.

The Coloured population was a general category devised to encompass everyone not catered for in the above categories. Spatially it was concentrated in the western Cape

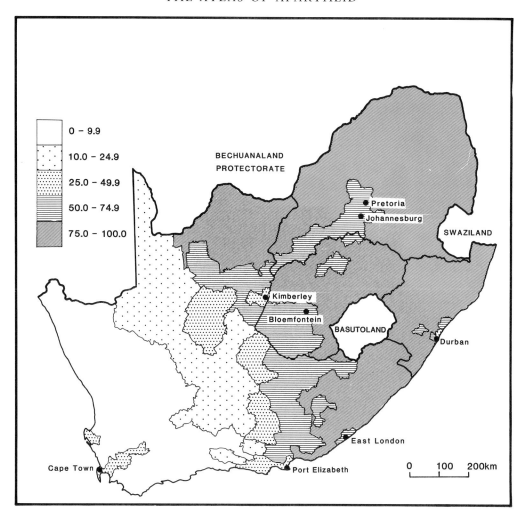

Figure 1.10 Percentage of population returned as Black, 1951

Source Computed from South Africa (1955) *Population Census 1951*, Pretoria: Government Printer

(Figure 1.12). The main component was made up of the people of mixed European ancestry, who were socially excluded from being accepted as European. Also included were the descendants of the slaves imported into the Cape Colony in the seventeenth and eighteenth centuries. They, the Cape Malays, were acculturated to the extent of adopting Afrikaans in speech, but spiritually found solace in Islam (Figure 1.13). On the basis of acculturation the descendants of the Khoisan population were also included, as they were incorporated linguistically into the Afrikaans speaking population. The Coloured community thus included a wide diversity of origins, which varied according to the region concerned.

22

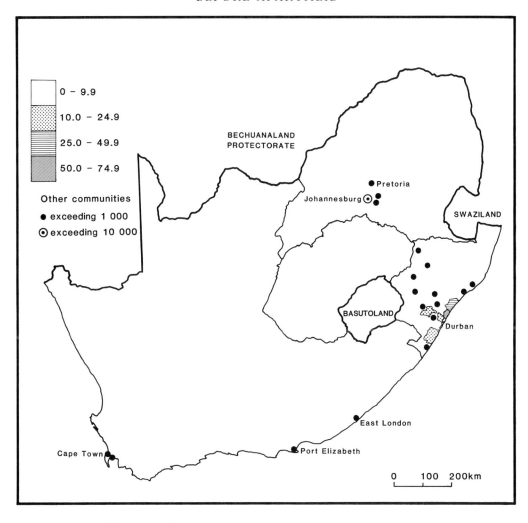

Figure 1.11 Percentage of population returned as Asian, 1951

Source Computed from South Africa (1955) *Population Census 1951*, Pretoria: Government Printer

THE ECONOMY

The colonial economy was dominated by the international market and its links to the interior of the subcontinent. Until the nineteenth century the value of South African imports and exports was extremely limited. Agricultural products, notably wine, wool and ostrich feathers, dominated the export trade until the 1870s (Figure 1.14). Thereafter minerals, first diamonds and then gold, assumed the dominance which they maintained for over a century. This is not to suggest that colonial society was completely commercially oriented and linked to the international economy. It has been

23

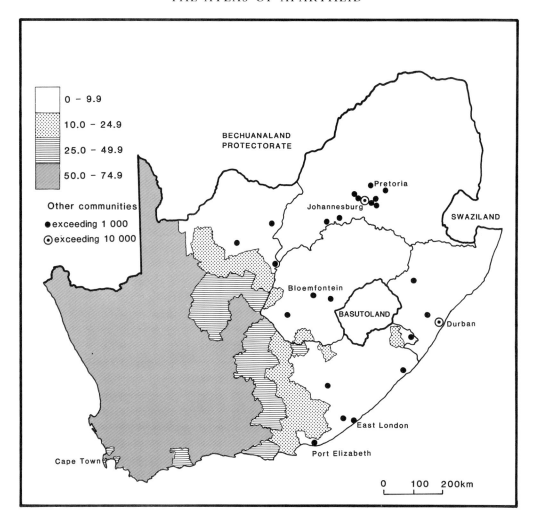

Figure 1.12 Percentage of population returned as Coloured, 1951

Source Computed from South Africa (1955) *Population Census 1951*, Pretoria: Government Printer

noted that parts of the White agricultural sector were only commercialized during the Second World War.

The agricultural sector played a vital role in the development of the country before 1948. The wealth generated by exports of wine and grain had given impetus to the emergence of a rural gentry by the mid-eighteenth century. In the nineteenth century the development of pastoral woolled sheep, angora goat and ostrich farming introduced commercialization to a wider extent of the country. In the twentieth century maize and sugar cultivation diversified exports and led to commercialization of more regions. Internal cattle and grain markets provided a further underpinning of the economy.

However, by 1910 agriculture only contributed 17 per cent of the national income, a figure reduced to 13 per cent forty years later (Figure 1.15).

The development of the mining industry in the last third of the nineteenth century transformed the national economy dramatically. Revenues generated by the mining of diamonds at Kimberley and gold on the Witwatersrand were invested in massive infrastructural and conspicuous spending. More particularly industrialization which had begun in the coastal port cities in the 1850s proceeded rapidly. Urbanization on a substantial scale was initiated with the demand for labour on the mines and the attendant services. Although initially much of the labour in mining was White, soon the heavy tasks were undertaken by Blacks. The growing demand for more labour, particularly on the gold mines, was such that mine owners had to resort to large-scale

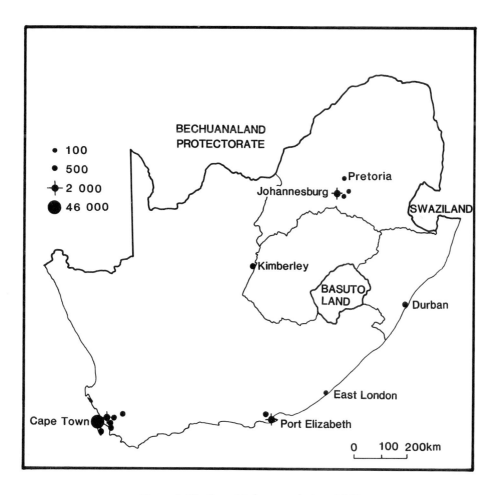

Figure 1.13 Cape Malay population, 1951

Source Computed from South Africa (1955) *Population Census 1951*, Pretoria: Government Printer

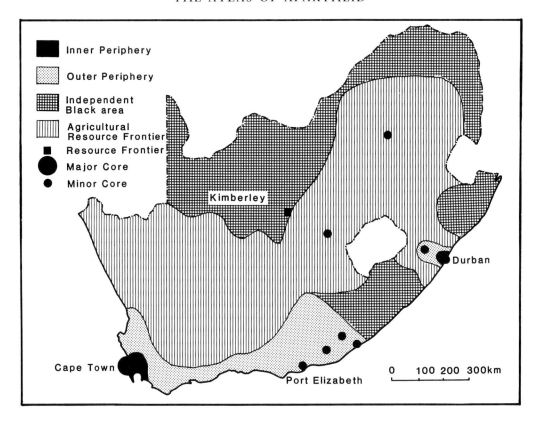

Figure 1.14 Development regions, 1870

Source After J.G. Browett (1976) 'The application of a spatial model to South Africa's development regions' in *South African Geographical Journal* 58: 118–29

recruitment from further and further afield. In the 1890s labourers were being recruited from neighbouring countries, including Mozambique. By the inter-war period all the countries south of the equator were represented in the South African mines.

The mining industry introduced large-scale migrant labour systems. Black miners were not recruited as permanent workers and therefore did not bring their families with them to live on the mining settlements. Instead the compound system, evolved for specific security purposes on the diamond mines at Kimberley, was adopted as the norm. Accordingly large hostels for Black adult males were constructed adjacent to the mine but separate from the remainder of the mining settlement for the permanent White (and Coloured) staff. The mining companies therefore did not have to bear the costs of supporting the families of the migrants, which were borne by the agricultural sector in the reserves.

Mining also transformed the space economy of the country. Both major discoveries were made on what were then peripheral parts of the settler economy. Kimberley was

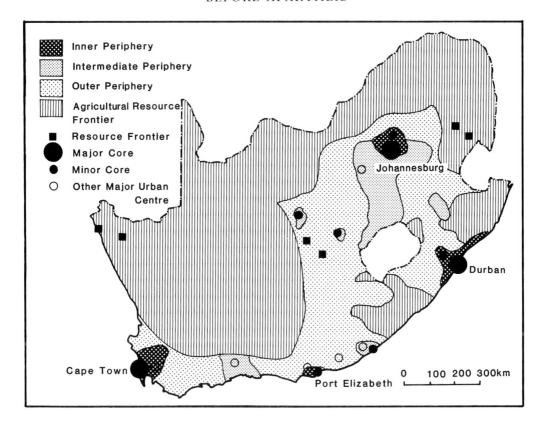

Figure 1.15 Development regions, 1911

Source After J.G. Browett (1976) 'The application of a spatial model to South Africa's development regions' in *South African Geographical Journal* 58: 118–29

situated on the then frontier of European settlement, but the diamond deposits were comparatively limited in extent. The Witwatersrand gold discoveries led to the establishment of a series of new towns extending from Krugersdorp to Springs, which became the hub of the national economy. The main settlement, Johannesburg, although only founded in 1887, was by 1911 the largest city in the country. The isolated coastal economies were then linked up to the Witwatersrand, which became the focus of the rail and road systems.

Attendant upon the development of mining was the expansion of basic industry and the manufacturing sector. At first agricultural processing had formed the basis of much local industry. To this was added basic metal work for the mining industry. This in turn led to the demand for the establishment of an iron and steel industry, which was only successfully undertaken by the State through the establishment of the South African Iron and Steel Corporation in 1928. State direction resulted in the selection of Pretoria as the site of the first plant and the careful integration of further plants within national

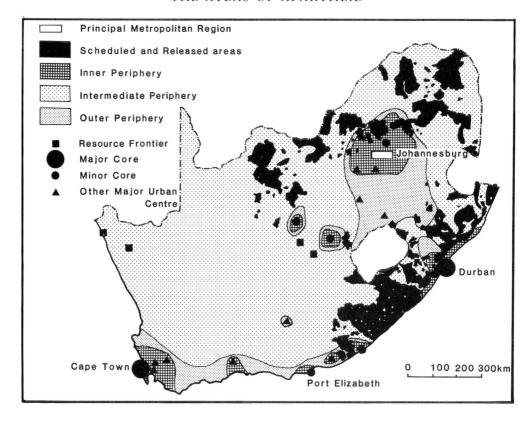

Figure 1.16 Development regions, 1936

Source After J.G. Browett (1976) 'The application of a spatial model to South Africa's development regions' in *South African Geographical Journal* 58: 118–29

planning policies. Manufacturing industry received a major boost in the Second World War when efforts to achieve a degree of independence from uncertain overseas suppliers resulted in the sector overtaking mining as the most significant contributor to the national income (Figure 1.16).

The boom in the economy associated with the Second World War resulted in a further impetus to urbanization. Black rural–urban migration was on such a scale that by 1946 there were more Blacks in the urban areas than Whites. The dominance of the metropolitan area extending from Pretoria to the Vaal River was well established by 1960 and accounted for nearly half the country's net output and employment in manufacturing (Figure 1.17). Such a concentration of economic opportunity attracted rural as well as urban migrants. The systems of control to prevent Black urbanization proved to be ineffective, while refusing to recognize the enormity of the problems raised by inadequate housing and services only resulted in confusion in official circles and extensive slum conditions. In 1948 a United Party government commission of enquiry

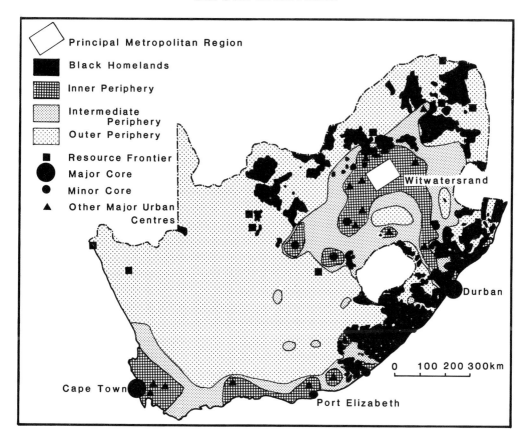

Figure 1.17 Development regions, 1960

Source After J.G. Browett (1976) 'The application of a spatial model to South Africa's development regions' in *South African Geographical Journal* 58: 118–29

recognized that the Black community was a permanent part of the urban population of South Africa. It further noted that the national economy was racially integrated and that the Black community formed an integral part of it. This report was in marked contrast to the prevailing concept that the towns were essentially White, both in control and occupation. Indeed a strong unifying political thread from the nineteenth century, in all parts of the country, had been the provision of separate urban locations or townships for the indigenous population, where social separation was re-enforced by physical separation.

THE POLITICAL DISPENSATION

The political development of South Africa prior to 1948 was complex. The Cape Colony was administered by the Dutch and later the British until 1910, although self-government was introduced in 1853. The republics established in the Transvaal and Orange Free State by the Afrikaner trekkers, escaping from British rule in the 1830s, were recognized as autonomous states by Great Britain in the 1850s. Natal, however, was annexed to the Crown in 1843. Later, ephemeral republics and colonies on the frontiers of European settlement were established, but by 1900 they had been incorporated into the four adjacent states. The anomalous survival of Lesotho and Swaziland as separate entities dates from this period of flux in colonial administrations, which became fossilized in 1910. Accordingly in 1910 the Union of South Africa was formed from the four provinces of the Cape, Natal, Transvaal and Orange Free State.

Control of the new British dominion was vested in a government elected by parliament. Parliament itself was controlled by the White population, through the exclusion of virtually all the remainder of the population from the franchise. Each province retained the form of franchise it had enjoyed in the colonial period. Thus in the Orange Free State and the Transvaal universal White adult male franchise endured. In the Cape Province and Natal a qualified franchise open to all was adopted, subject to property or income qualifications. In Natal the franchise had been restricted in such a way as to exclude the indigenous population and virtually the entire Indian population. In the Cape Province there was a much more open approach with a substantial Coloured and Black vote. Regarding the distribution of seats between the provinces, however, only White voters were counted constitutionally, leading to a built-in under-representation of the Cape Province throughout the Union period. In 1932 when women gained the vote it was restricted to Whites only, thereby halving the impact of the Cape Coloured vote. In 1936 Blacks were removed from the voters' roll and Senators appointed to oversee Black interests. Thus the South African state was firmly administered according to laws which served the interests of the White population.

Politically, through malapportionment and the gerrymandering of the colonial, nineteenth-century republican and Union electoral systems, the farming community dominated local parliaments until after the Nationalist victory in 1948. Thus policies demanded by the farming community were always sympathetically addressed by successive governments. Furthermore, the gerrymandering in favour of the rural areas ensured a disproportionate Afrikaner influence on parliament (Figure 1.18). Through much of the Union period the English voter had little influence on the course of government policy.

Between 1910 and 1948 there were only three prime ministers, all of whom had been Boer generals in the Anglo–Boer War of 1899–1902. They formed several administrations of various persuasions but with a consistent segregationist policy. The founding prime minister, Louis Botha (1910–19) and his successor, Jan Smuts (1919–23 and 1939–48) stood for reconciliation between Afrikaner and English speaking communities, while J.M.B. Hertzog (1923–39) was more concerned with developing a South African nationalism based on the concept, 'South Africa first'. All three were interventionist, placing considerable volumes of legislation on the statute book to regulate the relationship between the races, which each regarded as being one of the

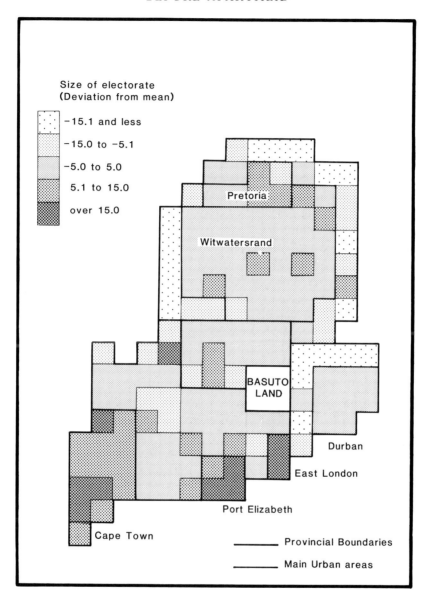

Figure 1.18 Malapportionment in parliament, 1948

Source Drawn from statistics in South Africa, Report fo the Ninth Delimitation Commission, *Government Gazette* 3931, 13 February 1948

major areas of government concern, as well as numerous laws having an indirect bearing upon the subject.

White political groupings tended to revolve around splits in the Afrikaner vote. The broad grouping which General Louis Botha led at Union split with General Hertzog's establishment of the National Party in 1913. In 1923 General Hertzog was able to form a government with the support of the Labour Party, which was particularly strong among English-speaking White mineworkers on the Witwatersrand. Racial protection in urban segregation and industrial job reservation for Whites provided a joint programme for the two parties. In 1933 the economic crisis led to a political accommodation between Generals Smuts and Hertzog to form the United Party. A splinter group of Nationalists under Dr D.F. Malan established a rival Purified (later Reunited) National Party and proceeded to gain Afrikaner support until eventually gaining power in the election of 1948. In 1939 General Hertzog resigned over the issue of entering the Second World War in support of Great Britain, which he opposed. General Smuts returned as prime minister to pursue the war and play a prominent role in international affairs.

Control of parliament was particularly important in South Africa as the economy was characterized by a high degree of state intervention. State ownership of the infrastructure, including the railways and harbours, resulted in an integrated transport policy. State control of key industries determined their location and hence a role in the urbanization process. Through active fiscal and rating policies, certain branches of local industry were fostered and agriculture protected. The Black labour force on both the farms and the mines was regulated and controlled by a complex series of laws restricting movement. Town planning legislation was used as an instrument of control to limit urban Black numbers and confine them to specific parts of the town, just as rural policies confined much of the Black population to specific reserves. The South African colonial and post-colonial economies and societies were thus highly regulated by the government. The apartheid era was accordingly a continuation of the State interventionist policies of the previous era. It was the scale of intervention and the ruthlessness with which it was undertaken which distinguished the post-1948 era from that which preceded it.

THE LAND ACTS

The division of rural land between the Black and White communities had largely been made by the time of Union in 1910. One of the first actions of the new government was to effect a legal division by describing those areas of the country which were assigned to the indigenous population. Under the Natives Land Act of 1913 some 8.9 million hectares were defined as Native Reserves (Figure 1.19). Furthermore, the Act prohibited the purchase of land by members of the Black community outside the scheduled reserves, except in the Cape Province. Provision was also made for the prevention of other means of independent Black access to land, through squatting, leasing, share-cropping or labour tenancy. The latter provision was resisted in many parts of the country by White landowners as they envisaged the system as the only means of overcoming a dearth of labour. However, measures directed against share-cropping and

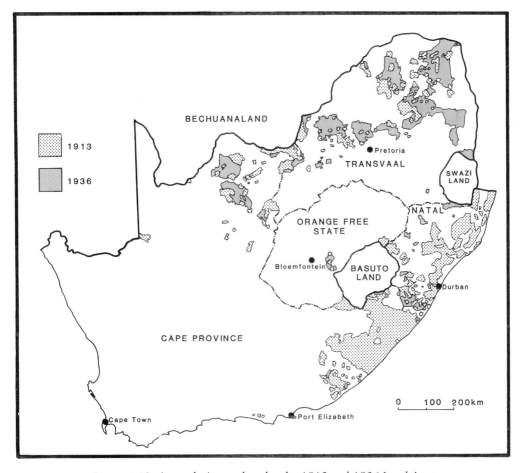

Figure 1.19 Areas designated under the 1913 and 1936 Land Acts

Source Redrawn from official 1:250,000 topo-cadastral maps, Government Printer, various dates

tenancy agreements were locally enforced, resulting in the initiation of the major resettlements associated with the removal of all Blacks not directly dependent upon White employers.

The Natives Land Act was officially conceived as a first stage in the drawing of a permanent line between the two races. A subsequent Commission of Enquiry (Beaumont Commission) reported in 1916 that extensive additional lands should be added to the reserves. Noting that some 3.6 million hectares of White-owned land and 0.8 million hectares of Crown land were exclusively occupied by Blacks, the Commission suggested they should be legally included in the Black areas, and additional provision made for those being displaced from White-owned farms. The First World War and the subsequent political turmoil delayed any legislative action on the recommendations for twenty years.

In 1936 the Native Trust and Land Act finally provided for the extension of the Black areas by some 6.2 million hectares. Here it should be noted that in most cases the schedule attached to the Act did no more than recognize the existing pattern of Black occupation, already noted by the Beaumont Commission. In terms of spatial patterns the broad consolidating additions suggested by the Beaumont Commission were rejected in favour of a more fragmented distribution to include individual farms occupied by Blacks (Figure 1.20). Furthermore, the implementation of the Act involved the purchase of the land through parliamentary grants, which was to be a slow process only completed more than forty years later. The 1936 Act also took away the right of Blacks to purchase land outside the Reserves and Trust Lands in the Cape Province and

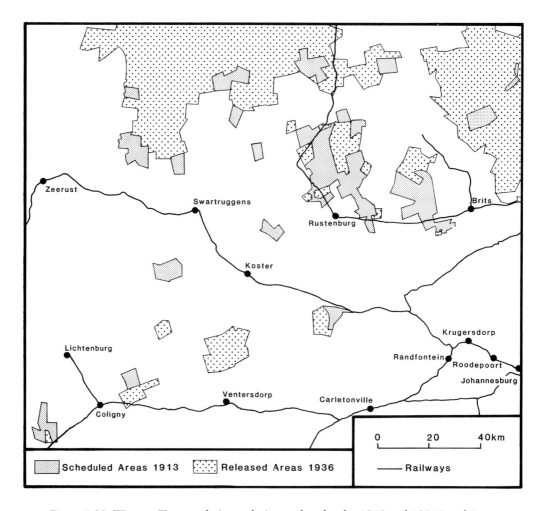

Figure 1.20 Western Transvaal: Areas designated under the 1913 and 1936 Land Acts

Source Redrawn from official 1:250,000 topo-cadastral maps, Government Printer, various dates

in a parallel act, Black voters were removed from the multi-racial Cape voters' roll. A South African Native Trust, subsequently renamed several times, was created to be the controlling agent of land purchased under the Act. Legislation was constantly amended to enforce the rules adopted to achieve the removal of Blacks from the White rural areas, except as paid employees, as the concept of segregationism became ever stronger in governmental thinking.

It should be noted that the Native Reserves and the Trust Lands accommodated less than half the total Black population during the Union period. In 1913 approximately half the Black rural population lived on the proclaimed Reserves and a further sixth on Mission stations, Black-owned farms, and lands exclusively occupied by Blacks. The proportion fell slowly in the Union period. In 1960 just under half the Black rural population lived on all categories of designated Black areas. However, as a result of the migration to the towns, only a third of the entire Black population continued to live in the Reserves and Trust Lands.

URBAN SEGREGATION

Formal urban residential segregation began in the aftermath of the abolition of slavery in 1834 and the first substantial influx of Blacks to the towns. Accordingly in 1834 the London Missionary Society in Port Elizabeth established a separate suburb for its charges, physically isolated from the remainder of the town. It was to be the first of several peripheral suburbs setting a precedent for the 'locations' which were to be established in South Africa in the ensuing 150 years (Figure 1.21). In 1847 the Cape colonial government promulgated regulations for the establishment of separate locations close to existing (White) towns where members of indigenous groups, including Afrikaans-speaking Coloured people, were required to live, if not housed by their employers or independent property owners themselves. The result was the establishment of subsidiary residential settlements throughout the eastern Cape in the second half of the nineteenth century. It should be noted that in the Cape Colony and Natal no racial restrictions were legally imposed on Black or Coloured ownership or occupation of land until the late-nineteenth century. However, few members of these two communities possessed the financial means to avail themselves of the legal opportunities.

The Orange Free State and Transvaal governments adopted the Eastern Cape system of separate residential areas for the Black and Coloured populations. The degree to which enforcement took place varied from strict regimes in the Orange Free State to relatively ineffective enforcement in the majority of Transvaal towns before 1900. The Natal authorities effected influx control measures to prevent a permanent Black presence in the towns, relying upon a migratory system to meet urban demands for labour.

The mining companies introduced segregated compounds to house Black single migrant workers in the settlements under their control. Strict supervision of such

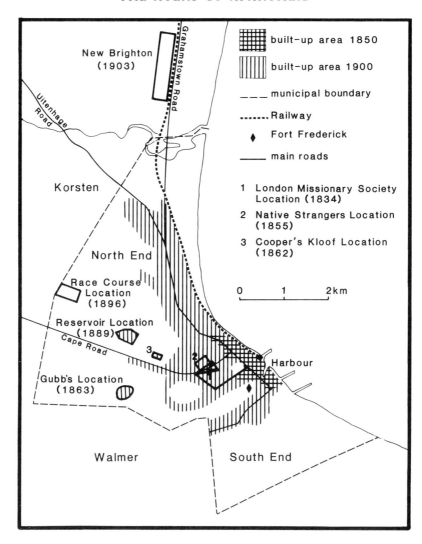

Figure 1.21 Black locations in Port Elizabeth, 1834–1903

Source After A.J. Christopher (1987) 'Race and residence in colonial Port Elizabeth', in *South African Geographical Journal* 69: 3–20

compounds was introduced to overcome the problem of illicit diamond smuggling in the Kimberley area (Figure 1.22). However, the degree of control which could thereby be exercised over the costs of labour and over the miners' activities made the system attractive for the subsequent gold- and coal-mine owners who did not have the same problem. Thus the steady decline in mining wages in the late nineteenth and early twentieth centuries has been attributed, in part, to the settlement form adopted to control the labour force.

At the time of Union a wide variety of urban policies relating to the indigenous population were in force, in contrast to the reasonably uniform rural reserve policies. The attempt to introduce national uniformity was only undertaken in the 1920s, as a result of the accelerated migration of Blacks into the towns and the consequent competition between Black and White workers. In 1922 the Transvaal Local Government Commission recommended a general tightening of regulations and the entrenchment

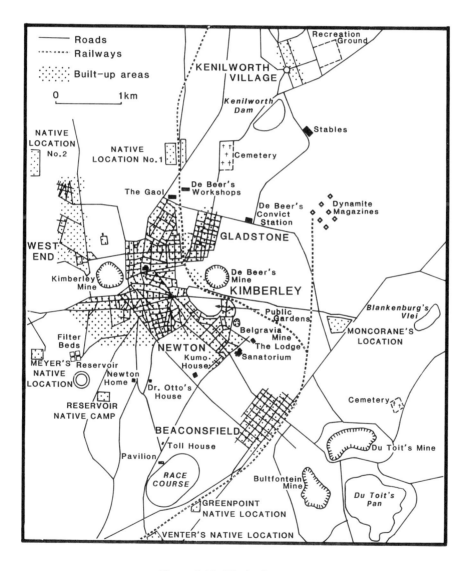

Figure 1.22 Kimberley, 1900

Source After A.J. Christopher (1976) *Southern Africa*, Folkestone: Dawson of the concept of the urban areas as the 'domain of the White Man' in which 'there was no place for the redundant Native'.[1]

In 1923 the Natives (Urban Areas) Act was passed requiring local authorities to establish separate locations for the Black population, and to exercise a measure of control over the migration of the Black population to the towns. The Act was amended and tightened regularly, notably in a major consolidating act in 1945.

A number of municipalities resisted the development of locations, questioning the expense involved. Durban municipality, for example, did not build its first location until the 1930s, preferring that employers, whether private individuals or companies, undertake the financial responsibility. Municipal councils and their predominantly White electorates were reluctant to spend funds from the general rate fund on the construction and administration of Black locations. The 'Durban system' of funding the construction of houses and services through the sale of indigenous beer was widely adopted, but severely limited the sums available for development. Hence, in contrast to the White suburbs, the Black areas were remarkably devoid of services, a legacy still in evidence.

By 1948 the system of physically separate Black locations was firmly in place throughout the country. Indeed the majority of towns had established Black locations by the time of Union, although those in the western Cape essentially housed the Coloured population at that date (Figure 1.23).[2] However, large numbers of urban Blacks lived elsewhere in the towns and cities, either in Black freehold properties, or as tenants in backyards. The continued housing of domestic servants throughout the White suburbs constituted a major feature of most towns.

Furthermore, those central city areas which were relatively integrated were made subject to the provisions of the Slums Act of 1934. This was applied for demolition of various inner but dilapidated suburbs, notably on the Witwatersrand. The displaced Black populations were largely rehoused in segregated mono-racial municipal housing estates on the urban periphery. The Slums Act was widely applied in the major metropolitan centres in order to impose racial segregation in a 'non-racial' manner.

The development of state housing schemes after the First World War was also a means of enforcing segregation. Under the 1920 Housing Act central government funds were made available to local governments to build housing for the poor. The resultant estates were required to be racially segregated, separated from one another by open spaces and with separate access roads. Although Black municipal estates tended to be grouped in formal locations, those for the Coloured and White populations tended to be dispersed throughout the poorer parts of towns, without any concept of broad urban divisions. Figure 1.26 (see p. 43) indicates the situation of such estates within the wider zoning practices of the private housing market.

The Indian community was also the target of various discriminatory measures from colonial times onwards. In the 1890s the Transvaal government sought to restrict the commercial activities of Indians. Separate 'bazaars' were established on the edges of existing towns where Indians were required to live and conduct their businesses. As in

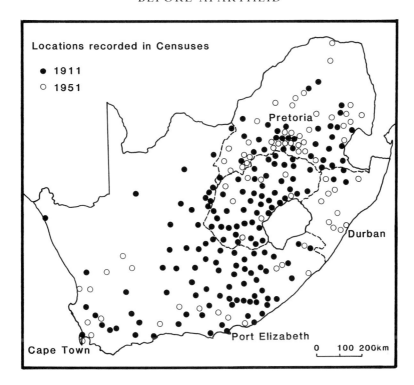

Figure 1.23 Black locations in South Africa, 1911–51

Source Compiled from manuscript location returns in the 1911 and 1951 censuses held in the Central Archives Depot, Pretoria, and the Central Statistical Services, Pretoria, respectively

the case of Black segregation measures, those applied to Indians were enforced with great variability, with few bazaars effectively occupied before 1900. The Asiatic Bazaar in Pretoria housed only 60 per cent of the Indian population of the city (Figure 1.24). It also included a separate 'Cape Location' for the Coloured population.

Anti-Indian agitation by the White population of Natal and the Transvaal resulted in an increasing level of restriction upon the residential options of the community throughout the Union period. In 1922 the Natal Provincial Council passed an Ordinance permitting the Durban Corporation to exclude Asians from accepted White suburbs. The policy was 'to separate the population of European descent, as far as possible, from Asiatics and Natives in residential areas – not to segregate any section or class entirely in parts of the Borough or from areas where at the present time any section has property or interests'.[3] The attempt to introduce national legislation under the Class Areas Bill of 1924 was withdrawn as a result of pressure from the Indian Imperial government.

In the 1930s attempts were made in the Transvaal to place limits on the extent of Asian occupation, through a detailed investigation by the Transvaal Asiatic Land

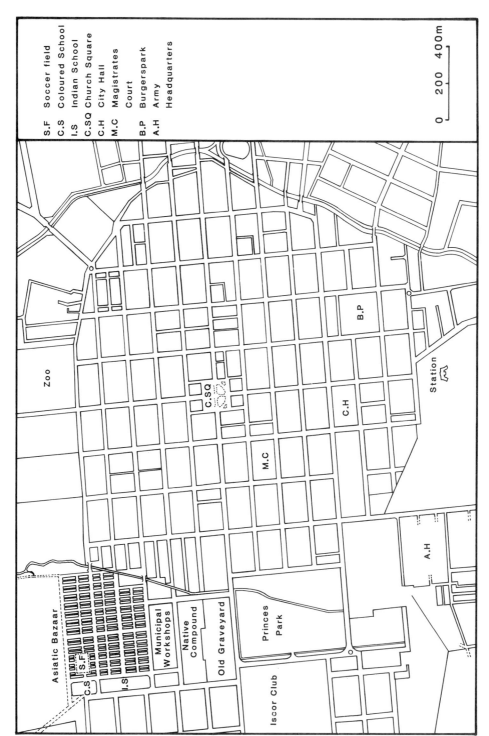

Legend:

S.F Soccer field
C.S Coloured School
I.S Indian School
C.SQ Church Square
C.H City Hall
M.C Magistrates
     Court
B.P Burgerspark
A.H Army
     Headquarters

0 200 400m

Figure 1.24 Asiatic Bazaar and central Pretoria, 1950

Source Taken from 1:10,000 map of the City of Pretoria (1950) Government Printer

Tenure Commission. Asian rights to acquire land remained tightly circumscribed, although through the employment of dummy companies and front men it was possible to avoid some of the impact of the law. At the same time the bazaar areas were occupied and the provisions of the Housing Act applied to the Indian population.

White complaints of Indian 'penetration' into predominantly White residential areas in the 1940s in Durban resulted in further government action (Figure 1.25). The enactment of the Trading and Occupation of Land (Transvaal and Natal) Act of 1943 and the more restrictive Asiatic Land Tenure and Representation Act of 1946 sought to confine Asian ownership and occupation of land to certain clearly defined areas of towns. This was to be achieved first through preventing inter-racial property transfers and then by establishing a Land Tenure Advisory Board to draw up plans for a permanent division of the cities into White and Indian sectors.

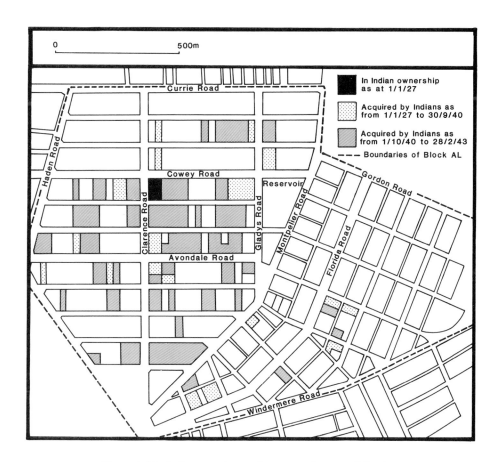

Figure 1.25 Indian purchase of property in central Durban

Source After South Africa (1943) *Report of the Second Indian Penetration (Durban) Commission,*
Pretoria: Government Printer

In the private housing market, township developers and some individuals introduced racially restrictive covenants into the title deeds. Usually these were designed to prevent anyone who was not regarded as White from owning or leasing the properties. The practice was begun in Natal in the 1890s as an attempt to prevent Indians from acquiring properties in the more expensive, and now literally 'exclusive', suburbs. Strict injunctions were often inserted into deeds of transfer, specifying those who were not wanted. A Durban deed for example forbade the ownership or occupation of land by any 'Arab, Malay, Chinaman, Coolie, Indian, Native or any other coloured person'.[4] However, private housing covenants included provision for resident Black and Coloured servants, who were exempted from the restrictions; 'only as such necessary domestics as the master or mistress of the dwelling house may require for his or her purpose in the comfortable habitation of the said dwelling house'.[5] By the 1930s few new private residential developments were laid out anywhere in the country without such restrictions. The extent of restrictive covenants may be gauged from the situation in Port Elizabeth in 1950 where most of the suburbs developed in the twentieth century were restricted to the ownership and occupation of a single group (Figure 1.26).

The range of residential options open to Coloured and Indian people were therefore progressively restricted, although not to the same extent as those open to Black people. Segregation within the urban areas of South Africa was accordingly well advanced by the time the National Party came to power in 1948, and was at a similar level to that experienced by the Black population in the United States at the same time. Only the Coloured and White communities in the western portion of the Cape Province experienced anything approaching an integrated residential pattern. Elsewhere racially defined segregation levels indicated the effects of long enforced and entrenched coercive measures.

LEVELS OF URBAN SEGREGATION

Despite the range of segregationary legislation and legal restrictions in operation after 1910, distinct regional patterns of enforcement and effectiveness are evident. The colonial heritage of regional segregation practices remained until the impact of the era of apartheid in the 1950s. Thus the White population, for whom the whole exercise was designed, exhibited markedly different patterns of urban segregation in the different provinces at the time of the 1951 census. If the standard index of segregation, measured on a scale from 0 (completely integrated) to 100 (completely segregated), is examined, the populations of South African towns and cities were markedly segregated by 1951, often as a result of the legal structures outlined.[6]

The residential separation of the White urban populations was most noticeable in all four provinces (Figure 1.27). The median national index of segregation for the White community registered 74.8. There were remarkable regional variations with significantly higher indices in the Eastern Cape and Orange Free State and lower in Natal. These reflect the contrasting colonial heritages of the different regions of the country.

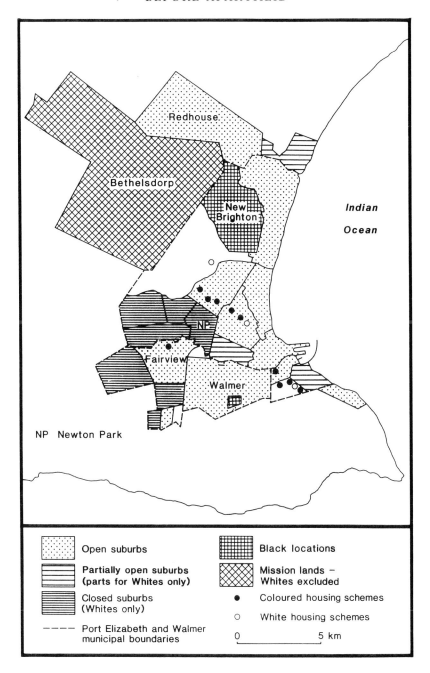

Figure 1.26 Extent of zoning by title deed in Port Elizabeth

Source Compiled by the author from Port Elizabeth municipal records and the Deeds Office, Cape Town

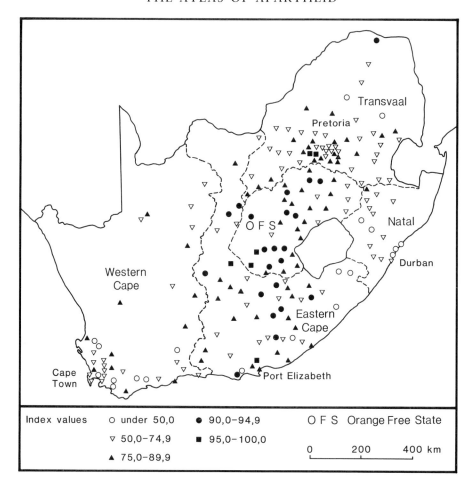

Figure 1.27 White index of segregation, 1951

Source Compiled by the author from the manuscript enumerators' returns of the 1951 census held by the
Central Statistical Services, Pretoria

An apparent anomaly in the figures is the low level of Transvaal segregation. In many
respects this was due to the incorporation of extensive smallholding zones within the
municipal boundaries, where large Black labour forces were housed on what amounted
to small farms owned by White suburbanites. Again this continued the nineteenth-
century colonial pattern where the majority of towns incorporated a substantial
agricultural component.

The national level of Coloured segregation indices was substantially lower with a
median value of 49.3 in 1951 (Figure 1.28). Regional variations were again extremely
marked. In the Orange Free State the Coloured population had generally been housed
with the Black community in the locations and therefore surprisingly registered the

44

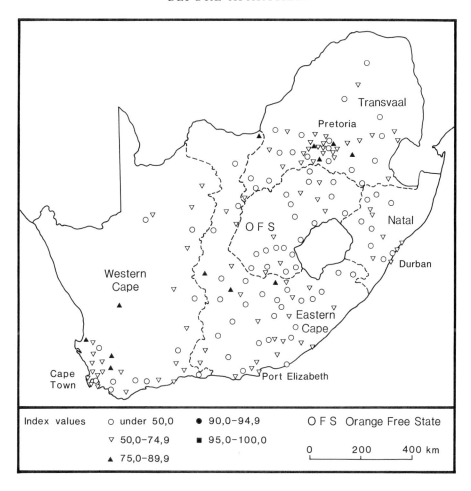

Figure 1.28 Coloured index of segregation, 1951

Source Compiled by the author from the manuscript enumerators' returns of the 1951 census held by the Central Statistical Services, Pretoria

lowest indices. In contrast the small Coloured community in Natal had for most purposes been integrated with the White population. The majority of the Coloured population of the country lived in the Western Cape where the majority of towns exhibited only moderate levels of segregation with a median index of 55.8. In a number of studies of American cities an index value of 60.0 has been taken to indicate a measure of structural segregation. In 1936 the median index had stood at only 44.3, indicative of the impact of segregated housing policies in the fifteen intervening years.

Asian levels of segregation exhibited a substantial variation in values (Figure 1.29). Even the high Transvaal median value of 80.1 masks a wide range from 32.5 to 96.6, reflecting the variety of municipal programmes directed towards the relocation of

45

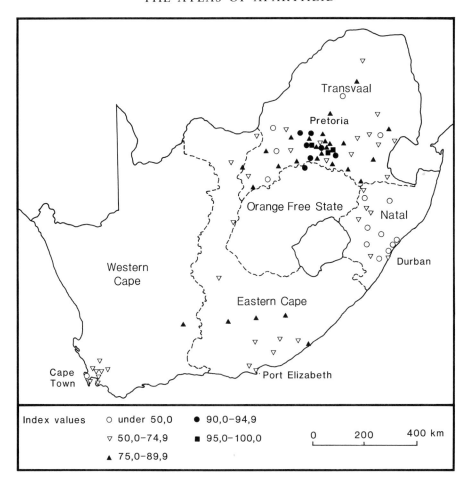

Figure 1.29 Asian index of segregation, 1951

Source Compiled by the author from the manuscript enumerators' returns of the 1951 census held by the Central Statistical Services, Pretoria

Indians in the bazaars and other segregation practices. Significantly, segregation levels between Asians and Blacks were higher than those between Asians and Whites as White urban administrations had made a concerted effort to exclude Asian retail stores from Black areas and thus enable White traders to retain part of the Black trade.

The Black indices of segregation exhibited a range of values corresponding in many ways to the White indices (Figure 1.30). In the Eastern Cape and Orange Free State high index values reflected the degree to which the Black population had been confined to locations or compounds. In these two regions over 70 per cent had been so restricted, a proportion little changed over the previous forty years. In Natal the slow progress of location building was apparent as only a third of urban Blacks lived in formal locations,

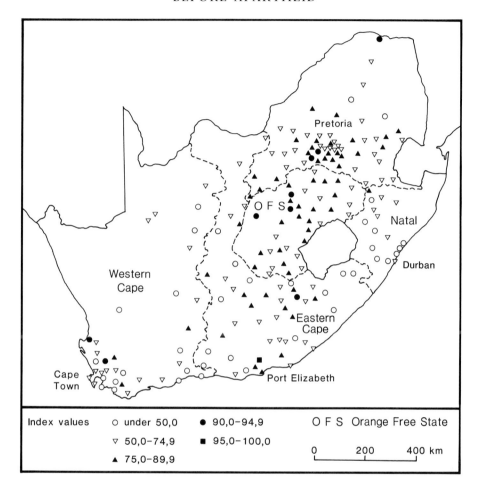

Figure 1.30 Black index of segregation, 1951

Source Compiled by the author from the manuscript enumerators' returns of the 1951 census held by the Central Statistical Services, Pretoria

compared with a national average of two-thirds. The remainder continued to live on their employers' properties, whether domestic or industrial.

It must therefore be emphasized that in the pre-apartheid era the majority of urban dwellers lived in segregated circumstances. In most cases this was the result of structural segregation practices ranging from title deed covenants to legislative prohibitions. However, the segregated suburbs were often separated by no more than a road resulting in greater contact than the statistics might suggest.

2

ADMINISTERING APARTHEID

The administrative machinery developed by the South African government between 1948 and 1991 was highly complex and expensive, involving the establishment of numerous overlapping jurisdictions and duplication of authorities. Some were inherited from the previous government, but the majority were new, and involved specifically with the implementation of the policy of apartheid. Some understanding of the administrative pattern is therefore necessary in order to follow the process and account for the regional variations in the outcome of government policies.

South Africa was established through the unification of four British colonies in 1910. Under the constitution little power was retained by the four provinces, and a highly centralized state came into being. Only such concerns as school education, health, environmental protection and local roads remained under provincial control. Some regional peculiarities, such as the divisional council system in the Cape Province survived until the 1980s with a measure of control by the local White electorate over social services in the predominantly rural areas. The various urban administrative systems provided for a greater measure of local control over municipal affairs. Even the spread of participation was wider as Coloured voters remained on the Cape municipal electoral rolls until 1972. Thereafter Coloured and Indian civic participation was limited to the introduction of Management Committees for the relevant group areas which could liaise with the White town council. A few Management Committees subsequently became fully fledged town councils, but this option was generally inoperable owing to the lack of financial resources and the Coloured and Indian communities' rejection of separation from the White municipality of which they had previously been a part.

Opposition to the central government's programme was thus only possible regionally from the provinces and the municipalities, where they were controlled by non-Nationalist councils. In the former case only the Natal Provincial Council was controlled by the opposition after 1948, although the Administrator was appointed by the central government. City and town councils were often ambiguous in their opposition to the government. Durban corporation assisted in the drawing up of the guidelines for group area determination, yet was anti-Nationalist in other aspects of policy; while the Cape Town City Council opposed all aspects of the implementation of the Group Areas legislation as far as possible, although limited by the control which the central government exercised over local government finance. The era though was one of increasing levels of centralization where elected provincial councils were finally abolished and municipalities lost much of the independence that had previously existed. In place, a series of centrally controlled bodies ensured the dominance of the central government in carrying out the official programmes of the state.

The central government ministries maintained regional offices, where necessary staffed by professional bureaucrats who were transferred from one part of the country to another in the course of their careers. The opportunities for the development of regional bias were therefore weak. However, local offices through the creation of pools of local expertise were able to modify the details of central government policies.

The most important local administrative unit of central government was the magisterial district. This was originally devised for the administration of justice and the collection of revenue in colonial times. Because of the significance of these functions, it was the maps showing magisterial districts which were most commonly used for administrative purposes and for the collation of statistics. The districts were based on urban spheres of influence until the 1950s, and generally disregarded the boundaries of Black land holdings (Figure 2.1). The standard topographic maps employed magisterial district boundaries and were therefore prominent in the cartographic presentation of state policies. In contrast, municipal boundaries did not appear on the national map series. District boundaries were not permanent and a constant creation of new districts makes the comparison of statistics remarkably difficult at anything other than national level. By the time of the 1985 census, district boundaries reflected Black state boundaries in a complex system of jurisdiction, which had substantially modified the map of magisterial districts in the eastern portion of the country (Figure 2.2). Even the boundaries of the four provinces were redrawn in the 1970s and 1980s as a result of the implementation of the state partition policy.

BLACK ADMINISTRATION

Parallel with the civil administration was that responsible for the Black population. The Department of Native Affairs and its successors operated a parallel administrative system through Native Commissioners and successor offices throughout the period from 1948 to 1991. The administrative apparatus combined with that of the magistracies in the areas covered by Native Reserves, but was run parallel to it in the remainder of the country. In 1948 there were six Chief Native Commissioners (later eight) and a substantial subordinate hierarchy (Figure 2.3). Notably the Chief Native Commissioner in Johannesburg also held the title Director of Native Labour, possibly an indication of the purposes of the entire regime. Regional offices of this department controlled most aspects of the lives of the Black populations, notably through the operation of the detested pass system until 1986. Passes amounted to internal passports

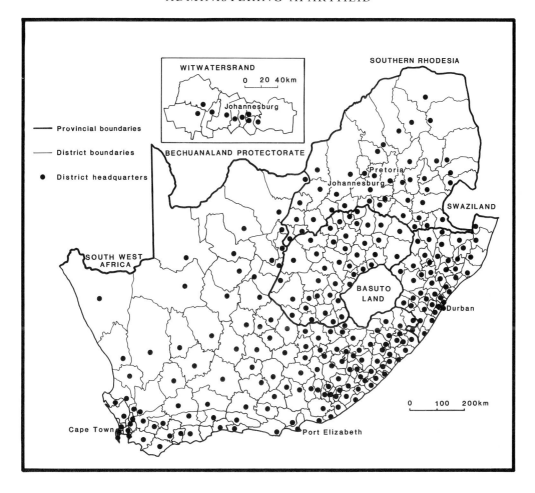

Figure 2.1 Magisterial districts and provinces, 1951

Source After South Africa (1955) *Population Census 1951*, Pretoria: Government Printer

allowing Black persons to move and reside in specified parts of the country. The acquisition of urban passes became progressively more difficult as the government imposed restrictions on permanent residence in the urban areas in favour of migrancy from the homelands. The later workings of this administrative system within the Black rural areas will be dealt with in Chapter 3 on State Partition. The administrative machinery was split between that applicable to the homelands and that responsible for the remainder of the country, under the Promotion of Bantu Self-government Act of 1959. In the former case Commissioners-General were appointed to oversee each of the Black ethnic groups recognized in the new organization.

A word on terminology is required at this point. Frequent changes in nomenclature afflicted virtually all government functions. Nowhere was this more evident than in

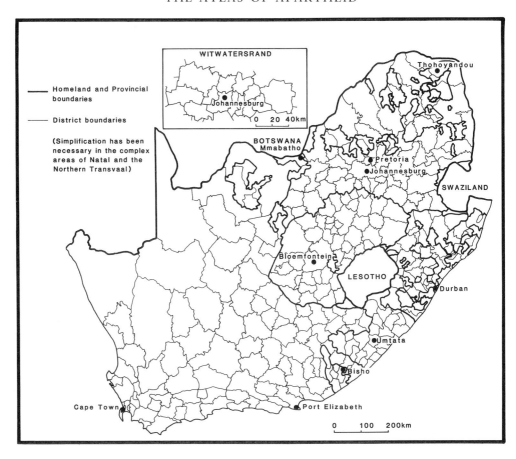

Figure 2.2 Magisterial districts and homelands, 1985

Source After South Africa (1986) *Population Census 1985*, Pretoria: Government Printer and maps of the four independent homelands

Black administration. In 1955 the Department of Native Affairs was divided into the Department of Bantu Administration and Development and the Department of Bantu Education (subsequently the Department of Education and Training). The former was renamed the Department of Plural Relations and Development in 1976, then the Department of Co-operation and Development, only to dissolve into a host of departments in the 1980s, including the Department of Development Aid, as functions were transferred to the homeland authorities. The various parallel organizations, except those relating to education, were finally merged with the national departments in 1991 as a part of the major reform programme introduced the previous year. The Department of Development Aid was wracked by a series of financial scandals relating to resettlement programmes. Townships consisting solely of hundreds or thousands of lavatories, but no houses or people, pointed to massive misappropriation of the taxpayers' money. The

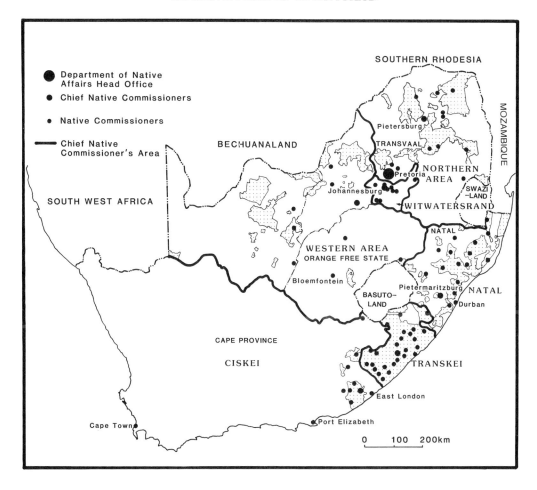

Figure 2.3 Native commissioners' offices, 1955

Source After South Africa (1955) *Summary of the Report of the Commission for the Socio-Economic Development of the Bantu Areas within the Union of South Africa*, Pretoria: Government Printer

'toilet town' of Restaurant in Lebowa became the ultimate *cause celebre* of a Department which lacked effective parliamentary oversight.

In the urban areas the separation of Black administration from that of the remainder of the city was effected through a series of measures. The Black Advisory Boards set up under the 1923 Natives (Urban Areas) Act were replaced in 1961 by Urban Bantu Councils. These remained largely advisory bodies and acted as agents of the local (White) authorities. In 1971 control of local Black affairs was transferred from the White local authorities to twenty-two Bantu Affairs Administration Boards, under the direct control of the Department of Bantu Administration and Development (Figure 2.4). After the 1976 Soweto riots the central government replaced the Urban Bantu Councils

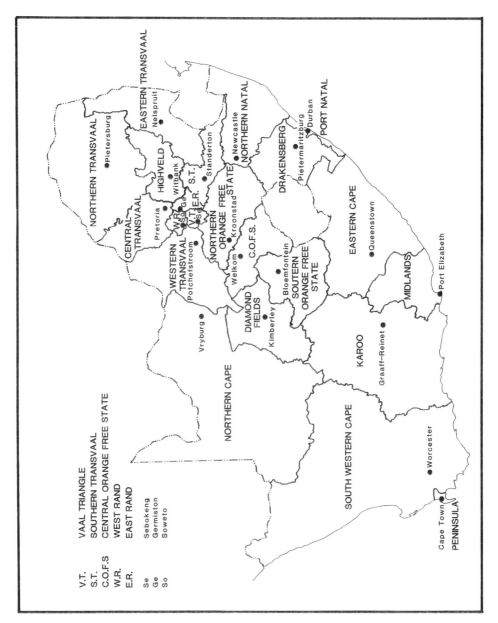

V.T.   VAAL TRIANGLE
S.T.   SOUTHERN TRANSVAAL
C.O.F.S CENTRAL ORANGE FREE STATE
W.R.   WEST RAND
E.R.   EAST RAND

Se   Sebokeng
Ge   Germiston
So   Soweto

Figure 2.4 Bantu Affairs Administration Boards, 1972–3

Source Based on information in the *Government Gazette*, various dates 1972–3

with Community Councils. Some 224 such councils had been established by 1980 but were unable to achieve any notable legitimacy with their constituents. In 1982 the Black Local Authorities Act provided for the establishment of a series of local government structures similar to those operating in the White areas. In the 1983 local authority elections only 7 per cent of the Black population cast votes for the new system, resulting again in a major problem of legitimacy. Thus many Black urban areas continued to be administered by White officials appointed to the task following the boycott of elections or the resignation or even murder of Black politicians.

REGIONAL PLANNING

Regionalization has been a constant theme as the central government sought to streamline the increasing complexity of local government structures. In 1982 the Department of Planning instituted a series of eight (later nine) planning regions for the country to co-ordinate national planning, notably that related to economic development (Figure 2.5). The regions were based on the major metropolitan regions, and incorporated the various Black homelands. This was in marked contrast to the majority of earlier national plans which had specifically excluded them. The move constituted one of the first major departures from the policy of state partition. Significantly, some homelands, notably Transkei, were divided between regions.

In 1985 the Regional Services Council Act provided for the establishment of joint metropolitan authorities for the distribution of bulk services. The first councils were established in 1987 in the major metropolitan areas and extended the following year to the rural areas. The councils were indirectly elected by the various local authorities represented on the council. However, the voting strength bore no relationship to population numbers, but was based on the financial contributions paid for the services consumed. Thus the Central Rand Regional Services Council had fourteen participating bodies, four Black, two Coloured, four Indian and four White. The voting strength of the White authorities totalled 74.48 per cent, while the four Black authorities constituted 19.84 per cent, although in population terms the proportions were reversed. The Regional Services Councils were thus constituted in a manner to ensure White control of this intermediate level of local government. A major exception to the Regional Services Council system was the Natal region where the KwaZulu government opposed its introduction, preferring a more equitable sharing of power between Black and White through the inclusion of homeland local authorities. In 1991 the government relented and Joint Services Boards were established for the Natal–KwaZulu region (Figure 2.6).

CONTROL OF PARLIAMENT

The most significant aspect of the political system was the monopoly of political power exercised by the National Party after 1948. In that year the Herenigde (Reunited) National Party and its ally the Afrikaner Party won a majority of the seats in the House

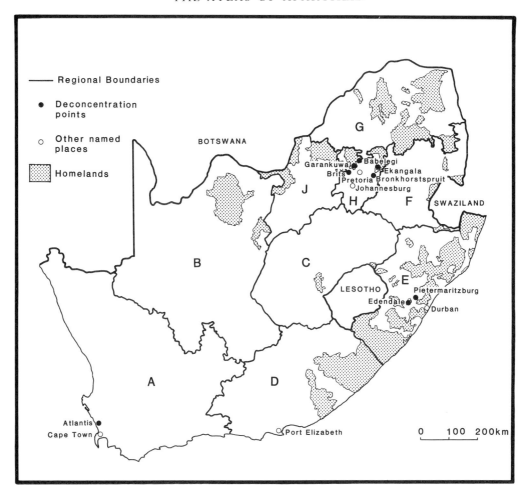

Figure 2.5 Planning regions, 1982

Source Based on Development Bank of Southern Africa map of South Africa

of Assembly, although only a minority of the vote (Figure 2.7). The element of gerry-mandering within the delimitation process was such that 40 per cent of the vote was translated into 53 per cent of the seats; even though the ruling, United, party had controlled the process and inadvertently assigned to itself larger than average constituencies and its opponents smaller than average. In 1951 the Herenigde National Party and the Afrikaner Party merged to form the National Party.

The delimitation process was then manipulated to ensure that rural voters retained a disproportionate voice in parliament (Figure 2.8). Until the 1980s the National Party was always assured of the majority of rural votes. The Union constitution provided for a 15 per cent range on either side of the provincial average constituency size. In 1963 it

Homelands excluded
P   Pretoria
W   West Rand
C   Central Rand
E   East Rand
V   Vaal Triangle
R   Rustenburg – Marico
H   Highveld
L   Lowveld – Escarpment
S   Stormberg
E.G East Griqualand

Figure 2.6 Regional and Joint Service Councils

Source Based on information supplied by the four provincial administrations

became possible for rural constituencies to be created which were 30 per cent smaller than the average, thereby packing the urban vote yet further. Such options were taken as injunctions to gerrymander by successive delimitation commissions. The result was to boost National Party representation in the House of Assembly. Thus in 1953 the party, although only securing an estimated 45 per cent of the votes, obtained 59 per cent of seats; while in 1966, the party's 59 per cent of the vote was translated into 75 per cent of the seats. The large number of unopposed seats resulting from the regional concentration of voting support, makes any discussion of the electoral system in South Africa remarkably speculative.

National Party support in the urban areas increased as Afrikaans-speaking Whites

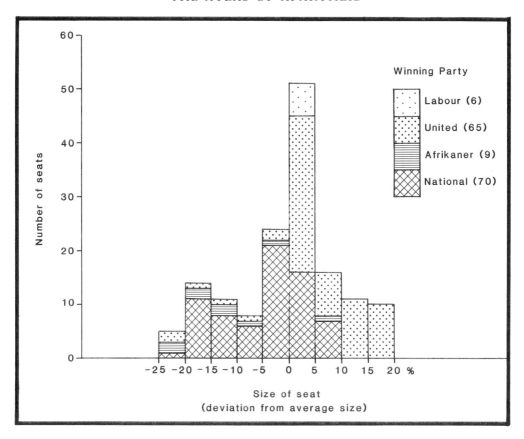

Figure 2.7 Size of constituency and winning party, 1948

Source Based on constituency sizes in *Government Gazette* No. 3931, 13 February 1948 and election returns in the *Eastern Province Herald* and *Oosterlig*, 27 May 1948

migrated to the towns. The government was also able to attract English-speaking support as a result of its policies. Further, the opposition United Party slowly disintegrated as prospects of regaining power receded. It finally collapsed in 1977, enabling the National Party to obtain an estimated 68 per cent of the vote and 82 per cent of the seats in that year's general election (Figure 2.9). Thereafter opposition from the Progressive Party and its successors on the left, and the Conservative Party, constituted in 1982, on the right, rarely presented any real threat to the continued supremacy of the National Party in the House of Assembly. The 1989 general election did however indicate National Party losses as the Conservative Party made inroads into the areas of traditional National Party support (Figure 2.10).

Parliamentary representation was also manipulated through the exclusion of the Coloured voters from the common voters' roll in the Cape Province in 1956. Exclusion only came after a five-year constitutional battle involving the packing of the Senate by

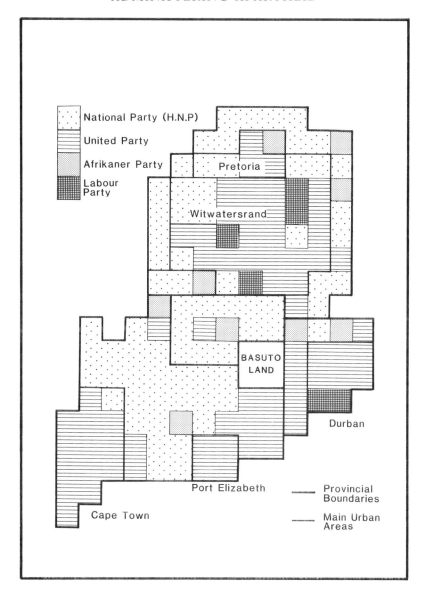

Figure 2.8 Election results, 1948

Source Based on constituency sizes in *Government Gazette* No. 3931, 13 February 1948 and election returns in the *Eastern Province Herlad* and *Oosterlig*, 27 May 1948

the National Party to ensure the overruling of the entrenched constitutional safeguards. At this time the Women's Defence of the Constitution League was founded, later to gain recognition for its human rights work as the Black Sash. The number of Coloured voters was comparatively small, but highly localized (Figure 2.11). The brief separate

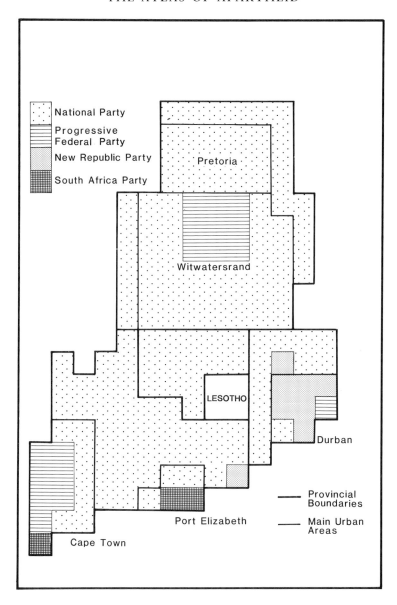

Figure 2.9 Election results, 1977

Source Based on election returns in the *Eastern Province Herald* and *Oosterlig*, 1 December 1977

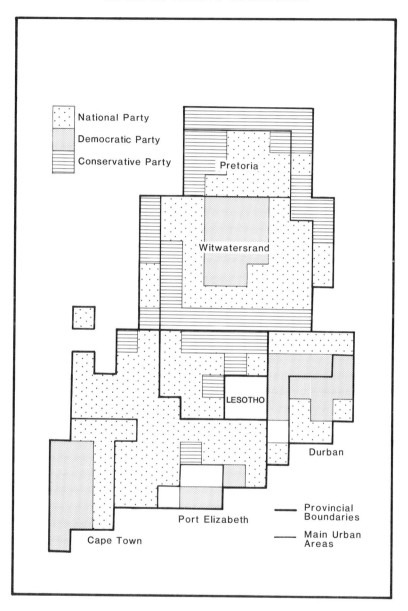

Figure 2.10 Election results, 1989

Source Based on election returns in the *Eastern Province Herald* and *Oosterlig*, 7 September 1989

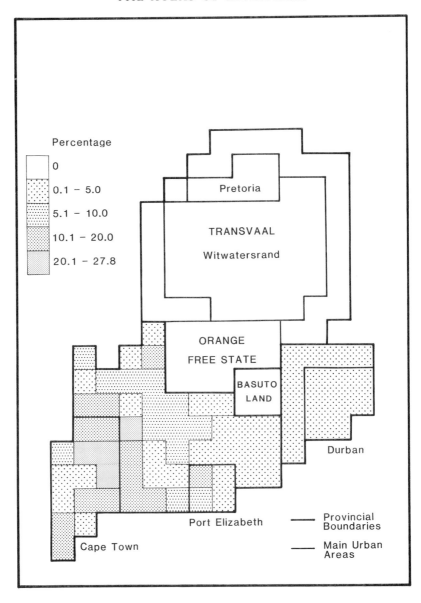

Figure 2.11 Percentage of electorate classified non-European, 1950

Source Based on statistics in South Africa (1953) *Official Yearbook of the Union of South Africa 1950*,
Pretoria: Government Printer

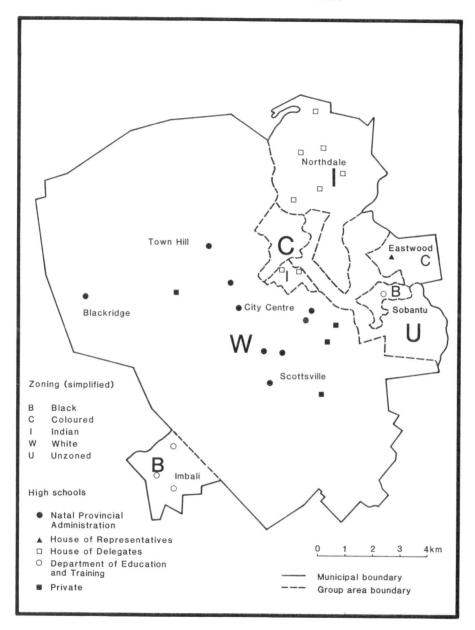

Zoning (simplified)

B   Black
C   Coloured
I   Indian
W White
U Unzoned

High schools

● Natal Provincial
   Administration
▲ House of Representatives
□ House of Delegates
○ Department of Education
   and Training
■ Private

——— Municipal boundary
– – – Group area boundary

0   1   2   3   4km

Figure 2.12 Controlling authorities for high schools in Pietermaritzburg, 1991

Source Based on information supplied by the relevant education authorities

representation of the Coloured community by four members of Parliament was an interlude before their total exclusion and the institution of the Coloured Representative Council in 1964. This was without any real powers and was abolished in 1973.

In 1984 a new constitution was introduced with a tri-cameral parliament. A House of Representatives for the Coloured community, and a House of Delegates for the Indian community were added to the, now all-White, House of Assembly to constitute the parliament. The balance of power between the three houses was heavily weighted in favour of the latter. The President's Council was established as an advisory body from the majority parties in the three houses, in which the ratio of members was 2:1:4. The President's Council could override the opposition of one or two houses of parliament, thus ensuring that the House of Assembly, dominated by the National Party, remained in ultimate control. Furthermore, the institution of an executive presidency effectively concentrated political power in the hands of the National Party.

ADMINISTRATIVE FRAGMENTATION

One of the features of the various constitutional changes in South Africa between 1948 and 1991 was the fragmentation of functions and the consequent multiplicity of authorities. Undoubtedly this provided a wide range of jobs within the civil service and hence a substantial interest group favouring the continuation of the system. Thus, for example, by 1991 there were seventeen separate departments responsible for state school education, namely the ten homeland authorities, the four provincial White authorities, those run by the Houses of Representatives and of Delegates, and the Department of Education and Training which was responsible for Black education outside the homelands. Accordingly in most towns and cities there were at least three if not more education authorities responsible for the state school system. Although mother tongue education was only enforced in Junior Primary schools in the Black education system, in the White system it was pursued throughout. Thus in Pietermaritzburg, as in most other towns and cities, the secondary system is even more confused, particularly as the private schools were largely denominational (Figure 2.12). The conflicting interests of the various education Departments inevitably led to crises in the implementation of Christian National Education and Bantu Education. Crisis was the constant state of affairs in the latter years of the apartheid era as Black school children formed one of the more visible groups opposed to the political system. Similar duplication was evident in many other areas of social service administration.

3

STATE APARTHEID

The philosophy of apartheid was introduced as the guiding principle of the National Party government under Dr D.F. Malan, which came to power in the general election of 1948, with the intention of creating a White Christian National State. Initially the new government differed only in the degree of enforcement from the segregationism practised by the previous administration. However, the guidelines were set in terms of the apartheid philosophy, which were to focus attention on South Africa in international affairs and give it its distinguishing features. Separation was to be introduced at all inter-personal levels ranging from separate park benches for Whites and other people to separate independent states for members of the various defined population groups. The principle of no equality between Black and White in church and state had been written into the Afrikaner republican constitutions of the Transvaal and Orange Free State in the nineteenth century, and the new government sought to emulate its predecessors in the mid-twentieth century.

State partition became the declared aim of the government as the policy unfolded. The major impetus to this policy came from Dr H.F. Verwoerd, who was appointed Minister of Native Affairs in 1950. In 1927 the Natives Administration Act effectively made the proclaimed Black areas subject to a separate political regime from the remainder of the country, ultimately subject only to rule by proclamation, not parliament. The Ministry of Native Affairs, as subsequently renamed several times, controlled a wide range of policies through what amounted to a parallel administration run by Native Commissioners. Dr Verwoerd used these powers to establish tighter control over the Black people both within what was perceived as the White part of South Africa and the various Native Reserves and Trust Lands. Separate government for the Black population was evolved into separate administrations, where the demands for political rights could be met without endangering White control over the remainder of the country. Thus state partition evolved as the cornerstone of apartheid to ensure continued White political control in South Africa in an era of Black emancipation on the remainder of the continent. It should be noted that it was only under the pressures of a changing international political environment that Dr Verwoerd accepted the full implications of state partition as the solution to White South Africans' numerical inferiority.

Under this policy all Blacks in South Africa would become members of a Black nation, which would possess a separate territorially based state, and within which the nation's political rights would be exclusively exercised. Thus at the end of the policy's implementation there would be no Black South Africans and the Whites, as the largest group of citizens, would be able to numerically dominate the government of the rump state. The possibilities of creating separate states for the Coloured and Indian communities were periodically discussed, but their establishment never became official policy. It was the extinction of the Black political presence in the country which was uppermost in the White policy makers' minds, as it was realized that Black numerical superiority constituted the greatest threat to continued White control. In practical terms this perception was also to be translated into a programme for the removal of as many Blacks from the White zone of South Africa as possible, leaving behind only those considered essential for the running of the economy.

NEW NATIONS AS THE BASIS OF PARTITION

The first problem which the government confronted was the need to create Black nations as the basis of nation-states. The government did not envisage the division of the country into Black and non-Black halves, where the Black zone would be numerically and economically powerful. In a classic example of the principle of 'divide and rule', the Black population was fragmented into a series of linguistically defined groups, which were the basic building blocks of the proposed dispensation. The broad Black or Bantu linguistic families were fragmented into ten subdivisions or 'national units'. Thus the Sotho linguistic family was divided into North, South, and West Sotho, or Pedi, Basutho, and Tswana respectively. Similarly the Nguni linguistic family was divided into Xhosa, Zulu, Swazi, Shangaan, North Ndebele, and South Ndebele. To these was added the Vendans of the northern Transvaal. It should be noted that in the 1950s none of these groups was significantly larger than the total White population, which was conveniently regarded as a unity, despite its linguistic and cultural diversity and political antagonisms.

TERRITORIES FOR NATIONS

The broad geographical distribution of these 'national units' gave some degree of order to the next stage of development (Figure 3.1). In 1959 the government enacted the Promotion of Bantu Self-Government Act, which sought to create a hierarchy of local governments for the Black rural reserves. Thus headmenships, chieftaincies, paramount chieftaincies, and territorial authorities assumed a place in an orderly progression of power in the Black areas. The Black population was accordingly assigned what was regarded as the group's 'traditional territory', even though the distribution reflected only the nineteenth- and early twentieth-century reserves.

The territorial basis of the system was the land designated as Native Reserves and

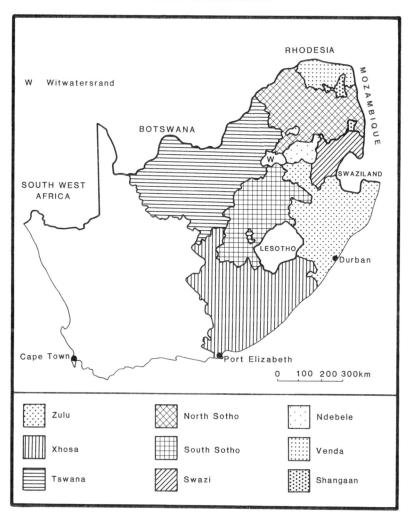

Figure 3.1 Ethnic division of the Black population, 1970

Source Based on statistics in South Africa (1973) *Population Census 1970*, Pretoria: Government Printer

that belonging to the South African Native Trust Lands. It needs to be remembered that the reserve boundaries were drawn according to the priorities of the colonial and other state governments, which certainly envisaged no form of separate sovereign existence for them. They were conceived as labour pools or homes for surplus Black people, but within the South African political arena. They were thus remarkably fragmented with few consolidated blocks of scheduled land to provide any respectable territorial basis for the partition. The Tomlinson Commission, which laid the framework for the social and economic development of the homelands in the 1950s, prompted the South African

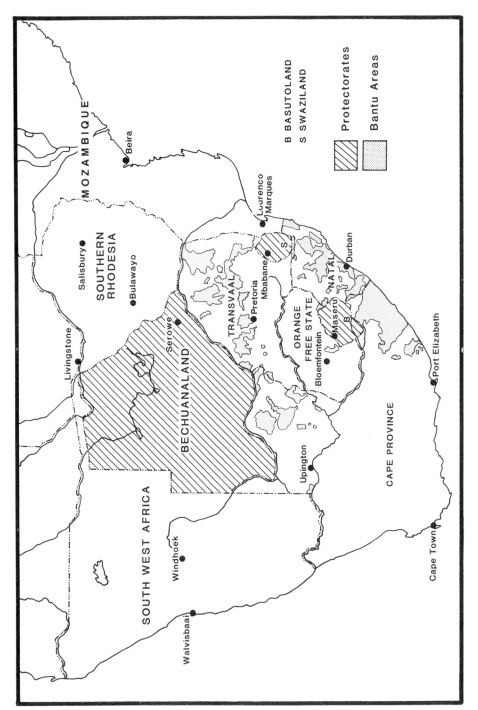

Figure 3.2 Tomlinson Commission map of South Africa

Source After South Africa (1955) *Summary of the Report of the Commission for the Socio-Economic Development of the Bantu Areas within the Areas within the Union of South Africa*, Pretoria: Government Printer

government to make further attempts to incorporate Basutoland, Swaziland, and the Bechuanaland Protectorate into the Union. They would then form the basis of three Black states to disguise the small proportion (13 per cent) of the country being apportioned to the new Black state structures (Figure 3.2).

The various Native Reserves and blocks of Trust Land were grouped together on a linguistic basis and brought under the supervision of a Chief Commissioner. The exception to this was the Xhosa people who were divided on historic grounds by the Kei River. Thus two Xhosa administrations evolved, one for the Ciskei and the other for the Transkei. Only in the latter had there been any system of consolidated local self-government before 1948. In 1966 the Ciskei government investigated the question of union with the Transkei, but rejected the option fearing loss of power to its larger neighbour. The other exception was the lack of any territorial base for the two smallest incipient nations, North and South Ndebele, until the 1970s. The administrative pattern which emerged was one of considerable intricacy. The Zulu lands, for example, were divided into over a hundred separate blocks when privately owned lands were included, suggesting comparison with the medieval German state system (Figure 3.3).

A second problem was the small proportion of the Black population living in the new states. In 1951 over 60 per cent of the Black population of South Africa lived in the areas designated as being part of the future 'White South Africa'. At state level, in virtually all cases, little more than half those claimed to be the citizens of the new states actually lived within their borders. As late as 1970 only 1.2 per cent of the South Sotho people of South Africa lived in the designated homeland of Qwaqwa, while 66.9 per cent of Vendans resided in Venda (Figure 3.4). Indeed in that year only 41.8 per cent of the Black population of the country lived in its designated state.

An added problem was the presence of ethnic minorities in virtually all the new states. Thus in 1970 only some 31.8 per cent of the Shangaan population of South Africa lived in Gazankulu, but another 21.5 per cent lived in other Black states. By judicial movements of population between the homelands and incorporation of strategic territory some 43.3 per cent of the Shangaan population of the country lived in Gazankulu in 1985. Minorities remained a source of friction between several of the new state governments, particularly where the basic ethnic identity of the population was disputed. Census enumerators in the Black states tended to find the characteristics of the dominant group rather than the minority while compiling the census data, thereby statistically reducing the extent of the minority problem. Homeland governments exerted pressures on minorities to conform to the language and culture of the governing group or face consequences of dispossession and expulsion. In the process of nation building the power of the new states became increasingly important in the quest for conformity and, resistance to it, the cause of often violent discontent.

The establishment of an Ndebele state illustrates the problems of a lack of a 'traditional territory' for the group. Until 1975 the South Ndebele constituted a minority within portions of Bophuthatswana and Lebowa, when they were encouraged to secede and form a separate state. The complexity of the situation in KwaNdebele

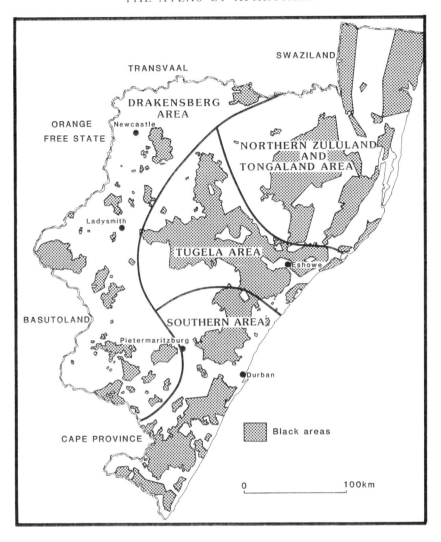

Figure 3.3 Fragmentation of Zulu lands in Natal, 1955

Source After South Africa (1955) *Summary of the Report of the Commission for the Socio-Economic Development of the Bantu Areas within the Union of South Africa*, Pretoria: Government Printer

was such that official plans were thwarted by the Bophuthatswana and Lebowa governments, which successfully sought to retain their territory in the region and expel the Ndebele inhabitants. Thus most of the state area had to be created from land previously in White hands, which was then populated by Ndebele displaced from other areas, notably Bophuthatswana (Figure 3.5). Furthermore, the attempt by the KwaNdebele government to opt for independence in the mid-1980s was thwarted by violent internal

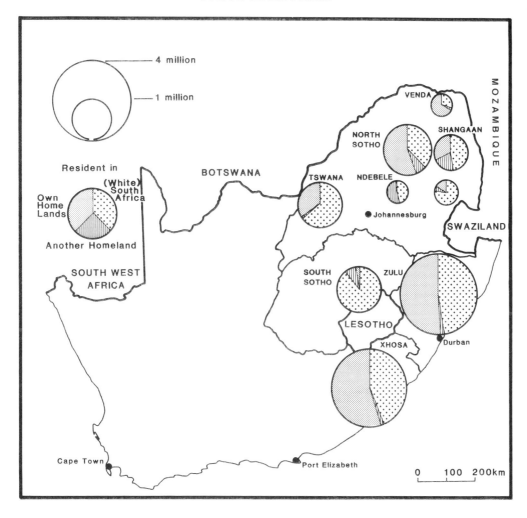

Figure 3.4 Proportion of Black groups in homelands, 1970

Source Based on statistics in South Africa (1973) *Population Census 1970*, Pretoria: Government Printer

dissension among a population consisting almost entirely of persons forcibly resettled from other parts of the Transvaal, including Bophuthatswana and Lebowa.

CONSOLIDATION

Territorial consolidation of the scattered reserves was suggested in the Tomlinson Commission Report in 1955. The suggested plans involved the High Commission Territories and a remarkably radical redrawing of the map of South Africa as the then existing areas were considered to offer 'no foundation for community growth' (Figure

71

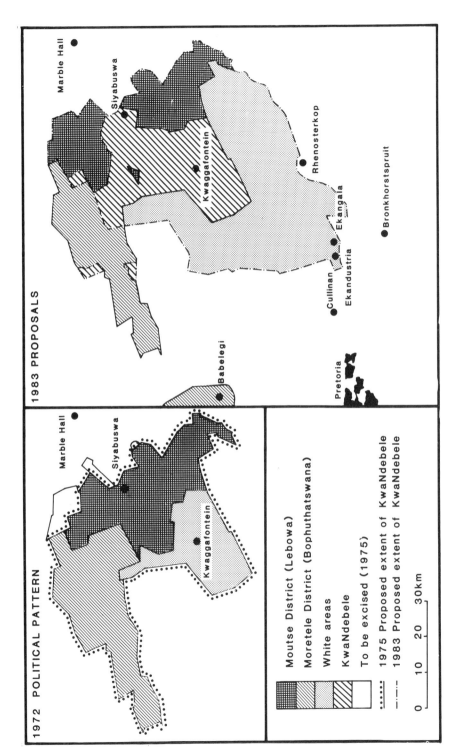

Figure 3.5 KwaNdebele

Source Modified after maps in L. Platzky and C. Walker (1985) *The Surplus People: Forced Removals in South Africa*, Johannesburg: Ravan

3.6).[1] The government rejected such expensive plans and proceeded with the consolidation of the Reserves and Trust Lands into larger blocks through a series of judicious land exchanges and purchases. Isolated tracts of Black land were excised from the homelands, while strategically placed land was purchased to join other detached portions to form larger blocks. In this manner the authorities expected to consolidate the Black territories into at most six portions per state. A severe limitation upon the programme was the political undertaking that the total areas of the new states would not exceed the area specified in the 1936 Native Trust and Land Act, namely 17 million hectares. It was only in the 1970s that this limit was reached, after forty years of restricted implementation, and the undertaking was then ignored. Clearly no broad sweeping consolidation was envisaged in official circles in the 1930s to create a viable territorial base for the new states.

Several consolidation plans were produced, each one more comprehensive than the last. The first comprehensive plan in 1973 involved a radical redrawing of the map of the Black areas of South Africa (Figure 3.7). Attention was directed towards those areas where most progress was made with the political programme of self-government and where Black co-operation was most evident and hence rewarded. This was reflected in the definitive 1975 plans which remained the basic framework thereafter (Figure 3.8). Thus the co-operative governments of Transkei and fragmented Ciskei received more attention than KwaZulu, where the Black authorities resisted government policy. In part this also reflected the initial adherence to the provincial distribution of areas authorized for incorporation under the 1936 legislation. There was no outstanding balance available in Natal. A further complication was the presence of over 800,000 hectares of Black privately owned land, which was also brought into the process of computing state areas. Accordingly, in Natal the scattered Black private farms were progressively expropriated and compensatory land adjacent to the Reserves acquired by the Development Trust. It should be noted that those farmers whose land was expropriated were entitled to no personal claim on the compensatory land from the homeland governments.

In the process of consolidation many of the smaller reserves and private farms were thus abolished. These areas were referred to as 'black spots' to be expunged from the map. In 1961 it was estimated that there were approximately 330 farms with a total area of 148,000 hectares falling within this category. The substantial change in the outlines of Bophuthatswana is a case in point, where the extensive purchase of land in the Transvaal before 1913 and the Cape Province until 1936 had resulted in the emergence of a scatter of Black privately owned lands in the south-western Transvaal and northern Cape (Figure 3.9).

The abolition of Native Reserve status frequently did not involve any form of compensation for the residents. The fate of the Mfengu Reserves in the Tsitsikamma Forest, some 150 kilometres west of Port Elizabeth, is an example where minimal compensation was paid for the houses, barns etc., but none was offered for the land. The blocks of land had been granted by the Crown to be held in trust for the various

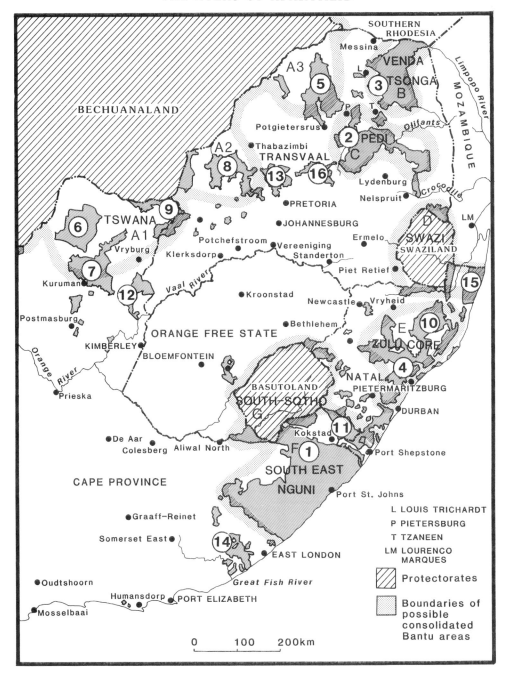

Figure 3.6 Tomlinson Commission consolidation plans

Source After South Africa (1955) *Summary of the Report of the Commission for the Socio-Economic Development of the Bantu Areas within the Union of South Africa*, Pretoria: Government Printer

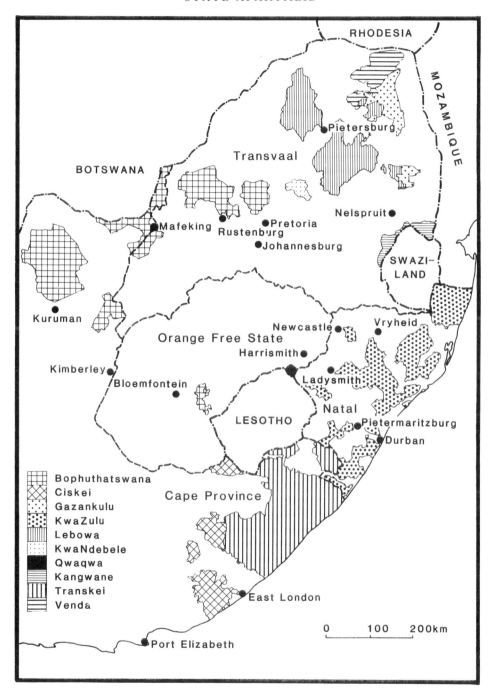

Figure 3.7 Homeland consolidation plans, 1973

Source After A. Lemon (1976) *Apartheid, A Geography of Separation*, Farnborough: Saxon House

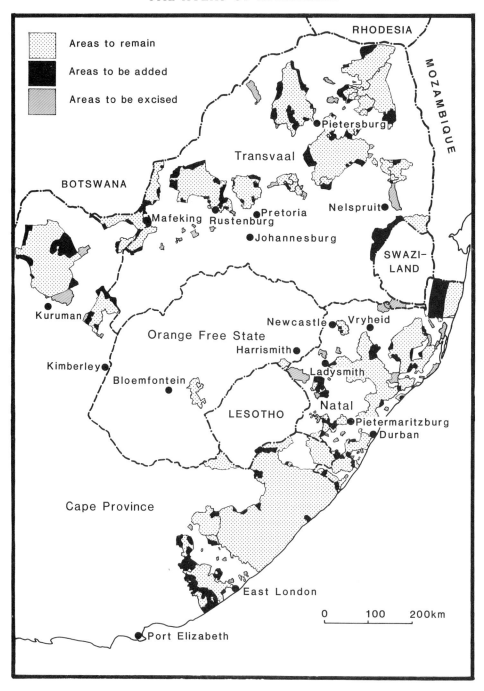

Figure 3.8 Homeland consolidation plans, 1975

Source After A. Lemon (1976) *Apartheid, A Geography of Separation*, Farnborough: Saxon House

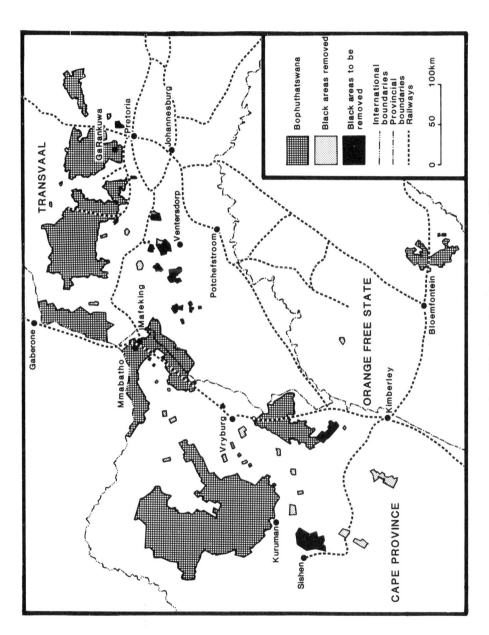

Figure 3.9 Consolidation of Bophuthatswana, 1972

Source After H.J. Moolman (1977) 'The creation of living space and homeland consolidation with reference to Bophuthatswana', *South African Journal of African Affairs*, 149–63

bands of Mfengu (Fingos) who had been displaced during the colonial frontier war of the early 1830s (Figure 3.10). In 1977–8 the community was moved 350 kilometres away to Keiskammahoek in Ciskei. Its members did not receive compensatory land as individuals had not held freehold title deeds to the land they worked. The area involved was included in the general compensatory land transferred directly to the Ciskei government, not to the people involved. The South African government subsequently divided the Tsitsikamma land into nineteen farms which were then sold to White farmers. The demand for the restitution of the land to the Mfengu makes the case a key issue in any discussion of land rights in the post-apartheid era.

The Tomlinson Commission had recommended the consolidation of the Black areas through the incorporation of adjacent South African Black territory into the three High Commission Territories. In 1982 the scheme was revived briefly with the attempt to transfer the Swazi national state area (Kangwane) and the district of Ingwavuma in northern KwaZulu to Swaziland (Figure 3.11). The latter piece of territory would have

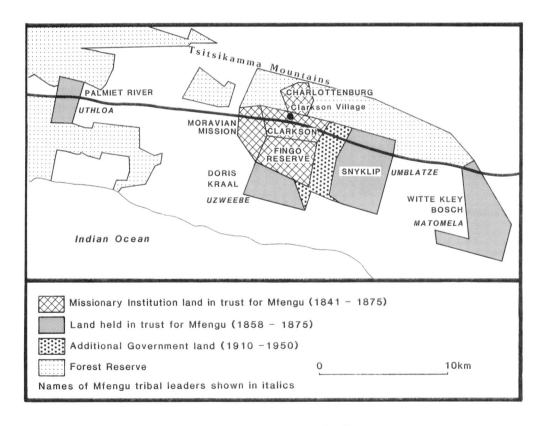

Figure 3.10 Mfengu Reserves, Tsitsikamma

Source Compiled from information extracted from the Deeds Office and Surveyor-General's Office, Cape Town

78

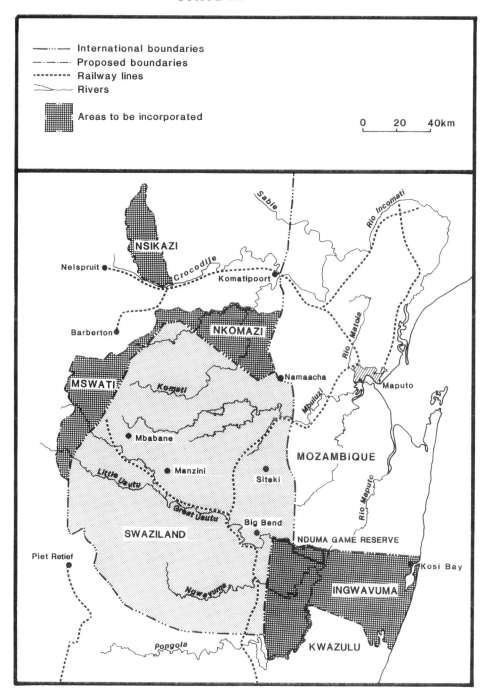

Figure 3.11 Swaziland extension proposals, 1982

Source After P. Esterhuysen (1982) 'Greater Swaziland?' *Africa Insight* 12: 181–8

provided Swaziland with a coastline, while the former would have partially overcome Swaziland's claims for restitution of territory lost to White colonists in the late-nineteenth century. The scheme was undermined by the opposition of the Kangwane and KwaZulu governments. The rectification of international boundaries which the plan involved incurred the opposition of the Organisation of African Unity, especially as it was being undertaken to foster apartheid. The scheme was abandoned with the death of King Sobhuza II of Swaziland.

RESETTLEMENT

The government also embarked upon a programme of resettling Black people from the White areas of the country in the homelands. The major movement was the resettlement of those considered surplus to the needs of the White farming community. The provisions of the 1913 Natives Land Act and subsequent measures were directed towards reducing Black occupation levels on White-owned lands. The practices of labour tenancy and sharecropping were also abolished. This led to a substantial reduction in the number of Black families living on White-owned farms, as both practices involved the presence of Blacks on White-owned land as semi-independent farmers, rather than as wage labourers. The continued presence of independent Black farmers on White-owned farms not occupied by their owners was also condemned and regulations enforced to end such occupation. A commission of enquiry was able to produce a number of maps which evoked marked political emotions (Figure 3.12).

The government accordingly sought two goals in the removal of Blacks to the homelands. The first was the aim of 'whitening' the rural areas of White South Africa. In 1921 Blacks, Coloureds and Asians outnumbered Whites in the White-owned rural areas by a ratio of 3:1. As a result of White rural depopulation, and the continued growth of the Black and Coloured rural population, by 1960 the ratio approached 9:1 (Figure 3.13). White complaints about the 'blackening' of the White rural areas, was a major spur to official action. The second goal was the converse increase in the number of Blacks living within their own designated national states and hence subjects of the homeland governments. Thus the proportion of Blacks living within the Black states had risen to over 60 per cent by 1985.

Estimates produced by the Surplus People Project in 1985 suggest that over 1.7 million people were displaced under this programme between 1960 and 1983 alone (Figure 3.14). The removals varied from family displacements by individual White farmers to large-scale forced removals organized and executed by the government, often with police coercion. Substantial movements took place particularly from the central Cape, which was perceived as a Coloured labour preference region (see Figure 4.17). The numbers were swollen by those displaced from the urban areas under the various influx control measures, who were also forced to move to the homelands.

Resettlement programmes in the Black states were of varying quality, often reflecting the nature of the displacement from the White farming areas. Frequently, no prior

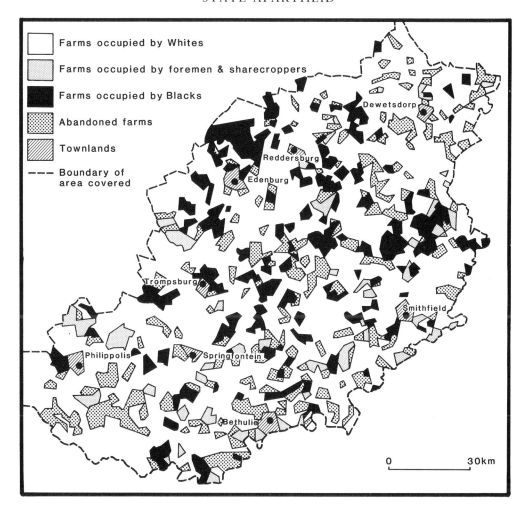

Figure 3.12 Black occupation of White farms, Southern Orange Free State, 1959

Source After South Africa (1960) *Report of the Commission of Inquiry into the European Occupancy of the Rural Areas*, Pretoria: Government Printer

preparation of reception sites was made and tents or other improvised shelters awaited those displaced. In other cases, resettlement villages were laid out and prepared with basic services before the arrival of the deportees of some of the larger organized programmes. Usually only sites were laid out before arrival. The result was a series of major resettlement camps such as Dimbaza in Ciskei, where Father Cosmas Desmond first drew the attention of the international community to the poor conditions in the camps and the plight of the refugees.

Resettlement was not a popular policy even for the Black homeland governments. Supplying the needs of the new arrivals placed considerable strains upon the limited

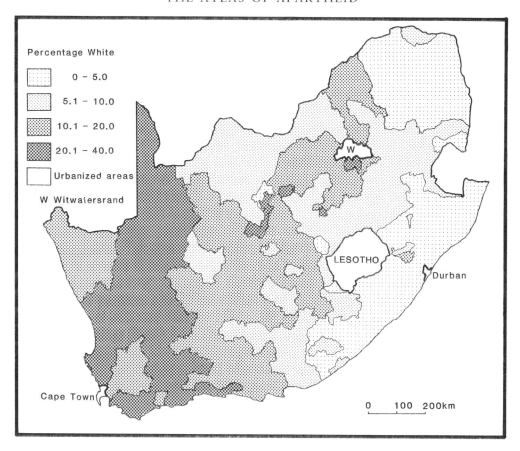

Figure 3.13 Percentage of rural population classified White, 1960

Source From statistics in South Africa (1963) *Population Census 1960*, Government Printer

resources of the administrations. Hence the South African government often chose to locate resettlement camps on Development Trust Land adjacent to the homelands and then negotiate their inclusion. Accordingly Ciskei was the scene of most Xhosa resettlements from the Cape Province as the Transkei government refused to co-operate and little Trust Land was incorporated into the state (Figure 3.15).

Furthermore, the new arrivals had no immediate place within the power structures which had been evolved in the early stages of the Black states' existence. Thus the refugees often remained among the poorest communities in the states and were generally without access to land either for grazing or cultivation. Also, few refugees were placed on the various agricultural schemes undertaken in the homelands which were reserved for those considered to be long-term citizens. Even those enticed to move from one homeland to another found the enticements impossible to gain in reality. The

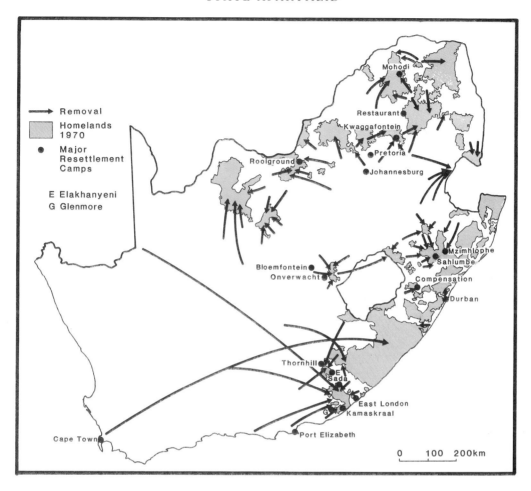

Figure 3.14 Forced removals

Source Compiled from information in the regional reports of the Surplus People Project

approximately 50,000 Herschel and Glen Grey refugees from the Transkei to the Ciskei in the post-1976 era fared little better than others displaced from White South Africa.

HOMELAND STRUCTURES

The homeland governments established their own administrations, which took over the extensive powers previously administered by the Native Affairs Commissioners and the Department of Native Affairs and Education. Executive authorities headed by Chief Ministers of self-governing states or Presidents of independent states were served by a range of ministries, which gradually assumed the running of the administration.

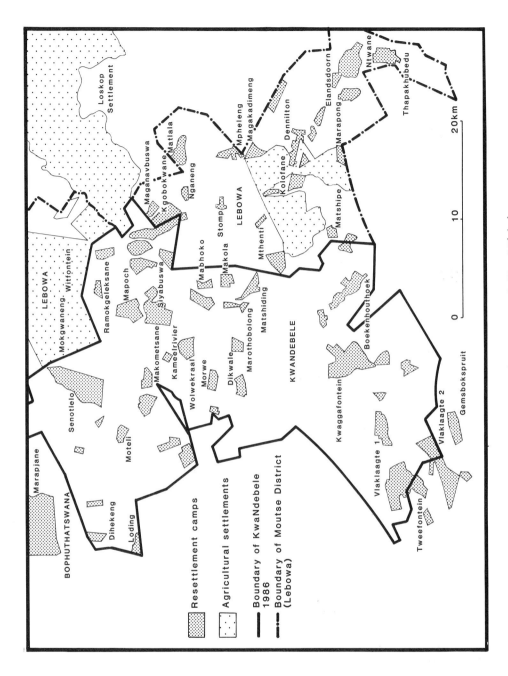

Key:
- Resettlement camps
- Agricultural settlements
- Boundary of KwaNdebele 1986
- Boundary of Moutse District (Lebowa)

Scale: 0   10   20km

Figure 3.15 Resettlement camps in KwaNdebele

Source Compiled from official 1:250,000 and 1:50,000 topo-cadastral maps

Bureaucracies provided significant areas of expenditure and employment for the Black population. Consequently, one of the immediate problems faced by all the homeland governments was the establishment of an administrative headquarters. As few towns were included within the homeland boundaries this constituted a major problem. Provisionally, the offices of the local Commissioner in the main White town were used. Then new offices and governmental complexes were built in the Black township attached to the major White town in the vicinity. In this manner Zwelitsha outside King William's Town and Seshego outside Pietersburg became the provisional capitals of Ciskei and Lebowa respectively.

The South African government drew up guidelines for the identification of the new permanent sites for the homeland administrations. They were basically the designation of an open waste area, removed from White influence, without historic significance, and which would be central to the future state area and population distribution. Further, it was suggested that the site lie on a development axis in terms of national development programmes. The nine homeland governments, joined by KwaNdebele in 1978, were then offered the official choices (Figure 3.16). It is worth noting that the requirement that the sites be without historic significance was rejected by several homeland governments once they gained sufficient power to influence state policies. Thus the South African government's choice of Nongoma for the KwaZulu capital was rejected in favour of historic Ulundi, site of the last royal capital of independent Zulu-land in 1879.

The central government's choice of Debe Nek as capital of Ciskei was rejected by the Ciskei government in favour of Alice, the seat of the University of Fort Hare and Lovedale College, the educational hearth of the Xhosa people. In other states only one suitable site was available. The restricted territorial extent of the Qwaqwa homeland was such that Witsieshoek, duly renamed Phuthaditjhaba, was the only site physically available. Most significantly Umtata was designated capital of Transkei, although nominally a White town, but entirely surrounded by Black reserves. This was to cause considerable controversy as other homeland governments claimed the major town associated with their state. Consequently, the Ciskeian demand for the incorporation of King William's Town was rejected by the South African government, although the Bophuthatswana government's demand for the incorporation of Mafeking was acceded to. As a compromise the Ciskeian government finally selected Bisho, adjacent to King William's Town, as its permanent capital site. It had little in common with any of the guidelines laid down by the South African government, being within 3 kilometres of a White town, was eccentric to the national territory and population and with a restricted site. It did serve notice, however, that King William's Town would be regarded as future Ciskeian territory (Figure 3.17).

The new state capitals were the subject of extensive conspicuous spending as the various governments established new complexes, official housing and the infrastructure commensurate with a state capital. In few cases was there an inherited infrastructure. New open sites offered considerable scope for grandiose town planning schemes. Thus

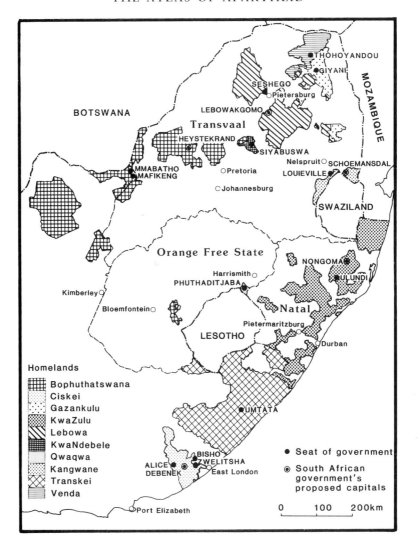

Figure 3.16 Homeland capitals

Source After A.C.B. Best and B.S. Young (1972) 'Capitals for the Homelands', *Journal for Geography* 3: 1043–55

Lebowakgomo, Mmabatho and Bisho were laid out with extensive government quarters, including the Presidential palace, parliament buildings, and government ministries. Expansive housing schemes supplied a range of housing styles and sizes previously not available for Blacks in South African towns, ranging from the Ministerial complex to civil service quarters and the latterly permitted squatter areas (Figure 3.18). Commercial and industrial sectors were also designated, in order to generate employment and promote the capital as the major town within the various homelands' urban hierarchy.

DECENTRALIZATION

Economic development was essential to support a greater proportion of the Black population within the homelands. The Tomlinson Commission had stressed the need for substantial investment in the homelands to prevent a massive flow of population to the (White) cities. However, the government showed little inclination to spend the sums required and only gradually developed any cohesive policy. A major philosophical problem had to be faced in the formulation of a homeland industrial policy. The possibility of White-owned capital being used to invest in industries in the homelands was rejected as it was considered that only Black-owned capital should be used in order to preserve Black independence. Little such capital was available in the 1950s and 1960s and hence few jobs could be generated.

At first restrictions were placed on the establishment or extension of industrial premises in the metropolitan regions (Figure 3.19). By 1971 the restrictions had been tightened so that no new industry in the Pretoria-Witwatersrand-Vaal Triangle region could employ more than two Blacks for every White employee.

Parallel with this policy was the border industry programme begun in 1960 under the auspices of the Permanent Committee for the Location of Industry, later (1971) renamed the Decentralisation Board. The object was to develop industrial areas in White South Africa, adjacent to the homelands, so that Black workers could commute on a daily basis between their homes and places of work. This programme overcame the ideological problem of White capital entering Black areas, and also meant that Black workers could live within the homelands.

Initially the Committee concentrated on those areas likely to attract industrialists in order to demonstrate the practicability of the policy, through the offer of financial and other incentives to establish factories in border cities. Three sites were developed, Hammarsdale, between Durban and Pietermaritzburg, Rosslyn, near Pretoria, and Pietermaritzburg itself. The Durban Metropolitan area was excluded as incentives were not required to attract industrialists to this region.

The initial phase was regarded as a success by the government, and in 1968 the geographical spread of sites was widened to include towns adjacent to other homelands. The range of incentives was also extended. The rate of new job creation, however, declined, suggesting that the more attractive sites possibly did not need to be subsidized. In the 1960s some 87,000 jobs were created in border areas, far short of the 20,000 per annum thought to be essential by the Tomlinson Commission in the early 1950s.

It should be noted that the numbers of cross-border commuters increased as the Black townships built in the homelands expanded when limits were placed on the growth of those within White South Africa. Accordingly Pretoria received an increasing

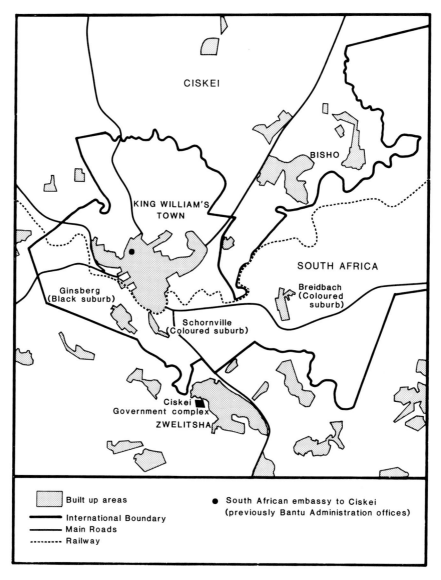

Figure 3.17 King William's Town and Bisho

Source Compiled from official 1:250,000 and 1:50,000 topo-cadastral maps

number of its Black workers from neighbouring Bophuthatswana, and subsequently also from KwaNdebele (Figure 3.20). Long distance commuting on a daily basis became a significant aspect of Black urban life as an increasing number of workers were forced to live in the homelands.

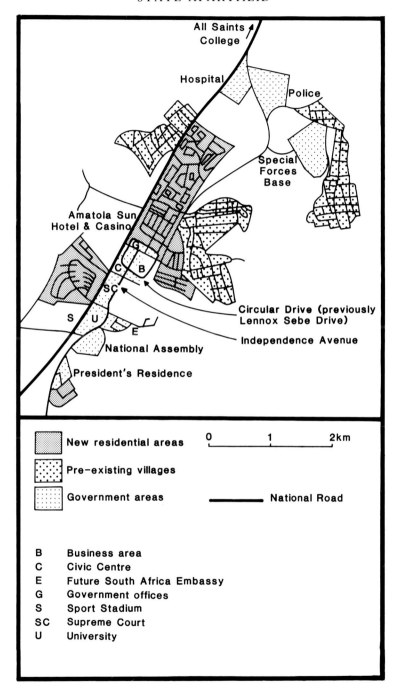

All Saints
College

Hospital

Police

Special
Forces
Base

Amatola Sun
Hotel & Casino

G

C   B

SC

Circular Drive (previously
Lennox Sebe Drive)

S   U

E

Independence Avenue

National Assembly

President's Residence

New residential areas

Pre-existing villages

Government areas

0   1   2km

National Road

B	Business area
C	Civic Centre
E	Future South Africa Embassy
G	Government offices
S	Sport Stadium
SC	Supreme Court
U	University

Figure 3.18 Plan of Bisho

Source Based on maps supplied by Bisho municipality

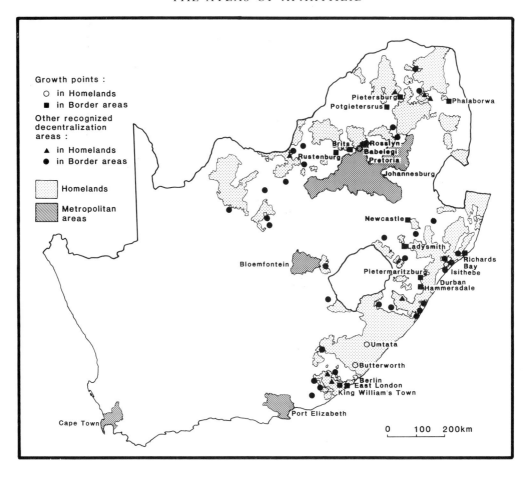

Figure 3.19 Decentralization policies, 1960–73

Source Statistics in T. Malan and P.S. Hattingh (1976) *Black Homelands in South Africa*, Pretoria: Africa Institute

HOMELAND GROWTH POINTS

In 1968 the government passed the Promotion of Economic Development of the Homelands Act, which provided for the controlled introduction of White capital into the homelands. Under this act it was possible for Whites to act as agents or contractors for the South African Bantu Trust, or the various homeland development corporations, providing the industrial or mining development became the property of the Trust or development corporation after a stipulated period of time (usually twenty-five years). With the philosophical objections to White capital entering Black areas overcome, the identification of growth points or industrial decentralization began. The industrial

90

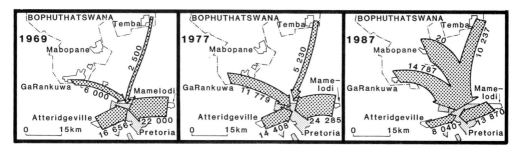

Figure 3.20 Changing patterns of daily Black commuter rail traffic in Pretoria, 1969–87

Source After A. Bernstein and J. McCarthy (1990) *Opening the Cities*, Durban: Indicator Project

decentralization programme underwent a number of changes in the course of the 1970s and 1980s, but the purpose remained the same. The scatter of development points was wide and inevitably only a minority were successful in attracting more than a few industries. The major growth points were in Bophuthatswana, notably at Babelegi, and in KwaZulu, notably at Isithebe (see Figure 3.19). Again, because of their proximity to the industrial heartlands of the Pretoria-Witwatersrand-Vaal Triangle (PWV) and Durban, this did little to decentralize South African industry. Other more remote centres such as Butterworth in Transkei, and Dimbaza in Ciskei were developed on the basis of generous government incentives. In a number of cases sites were only indicated vaguely on a map in order to provide each of the homelands with at least one decentralization point for political rather than economic reasons.

The incentives took the form of foundation grants, labour, power and transport subsidies and tax holidays, and so contributed to some measure of growth of employment. Long-term economic viability was rarely a consideration for many industrial concerns. The policy led to several unexpected results. In the southern Transvaal large numbers of workers were attracted to the points, but they were often not citizens of the homeland concerned. Thus Babelegi, although in Bophuthatswana, only housed a Tswana minority. Tensions arose in subsequent attempts to impose such rules as the use of Tswana as the medium of primary school education upon the non-Tswana population. The industrial development corporations also sought to attract overseas investment particularly from south-east Asia. Marked concentrations of Taiwanese and Hong Kong investments became evident. The capital remains essentially footloose, attracted by cheap labour and subsidies, and may not persist if those subsidies are withdrawn.

In 1975 the National Physical Development Plan sought to spread economic growth yet more widely with the introduction of development axes (Figure 3.21). New growth poles were identified to attract industries, and hence workers, away from the southern Transvaal metropolitan region. In 1982 the decentralization programmes were brought within a national framework of the revised National Development Plan. The multiplicity

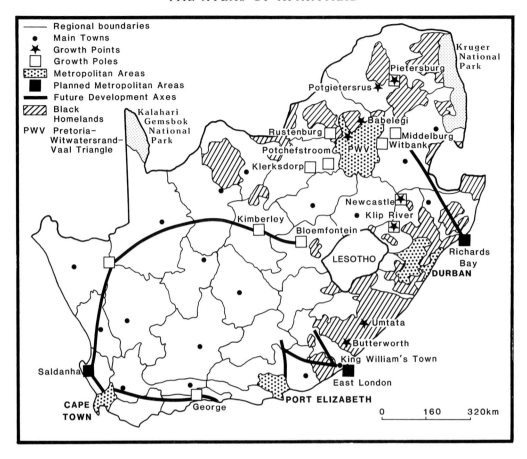

Figure 3.21 National Physical Development Plan, 1975

Source After T.J.D. Fair (1982) *South Africa: Spatial Frameworks for Development*, Cape Town: Juta

of development points remained, but attention was directed towards a few potentially more successful deconcentration points (Figure 3.22). Although a number of the deconcentration points were situated in the western parts of the country, as indeed had been some decentralization points, attention was primarily directed towards the homelands. Thus Ekangala, adjacent to the Ekandustria industrial area, was declared a deconcentration point in an effort to promote the development of the KwaNdebele homeland. The substantial financial incentives offered and the proximity to the industrial core region were such as to suggest, in the wilder official flights of fancy, a growth in population from 5,500 in 1985 to 750,000 within a few years. Similarly Botshabelo was designated as a non-Tswana deconcentration point primarily for the South Sotho population of Bloemfontein and those displaced from farms in the eastern Orange Free State. The designation of fewer points with greater incentives was expected to offer advantages, notably in an effort to boost economic growth rates in the economic heartland, which remains the driving force of the South African economy. The slowdown of the South African economy thus necessitated a review of the dispersed decentralization effort and its concentration in fewer, but more viable centres.

HOMELAND POVERTY

One of the most prominent aspects of the homelands has been their continuing poverty. Resettlement of unemployed people, high natural population growth rates, agricultural

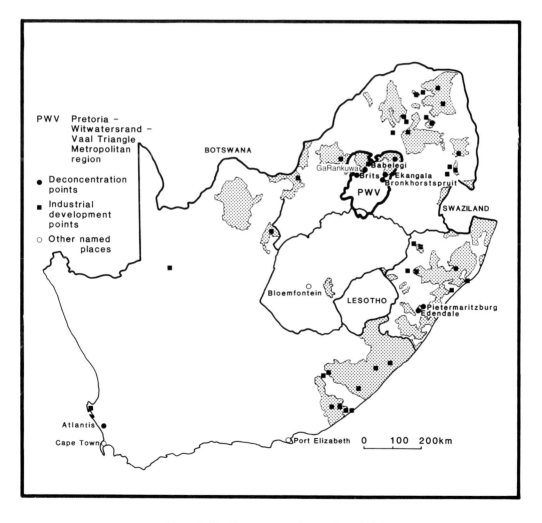

Figure 3.22 Deconcentration points, 1985

Source Based on map published by the Development Bank of Southern Africa stagnation and limited industrial development have all contributed to the perpetuation of a cycle of poverty. Without anywhere to go and without access to international capital on any scale, the majority of the homeland population exhibit remarkably low incomes, although some have prospered.

Estimates produced by the Development Bank of Southern Africa for 1985 indicated very substantial disparities in incomes. Whereas the Gross Domestic Product per head in the homelands varied from approximately R600–R150, that in the remainder of South Africa was approximately R7,500 per head (Figure 3.23). Only the development of mining in Bophuthatswana provided an income by the 1980s which indicated any degree of prosperity. Bophuthatswana also benefited from its proximity to the industrial heartland of the country, as the site of the major decentralization and deconcentration points. The disparities were narrowed through government subven-

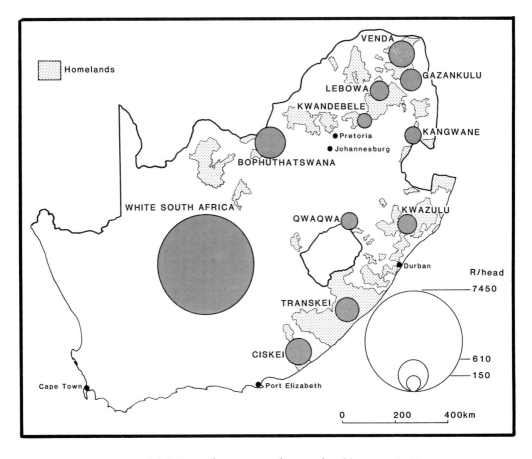

Figure 3.23 Gross domestic product per head by state, 1985

Source Based on statistics of the Development Bank of Southern Africa (1985) in G.M.E. Leistner (1987) 'Africa at a glance', *Africa Insight* 17: 1–104

tions to the homeland governments and the remittances of migrant workers, which accounted for up to three-quarters of Gross National Product. The relative position of the Black states had effectively deteriorated through the resettlement schemes and influx control, which prevented a poverty stricken rural population from migrating to the cities.

Nevertheless, one of the features of poverty was the high rates of temporary short-term labour migrancy from the homelands to White South Africa, which various official programmes of decentralization attempted to reduce. The impact of the migrant labour system affected the entire subcontinent (see Figure 7.13). However, within South Africa the effects upon the various homelands differed substantially (Figure 3.24). In 1970 only Gazankulu recorded a majority of males in the vital 15–64 age group temporarily absent from their homes. Generally those homelands furthest from the industrial areas recorded a higher ratio of absentees to residents. Thus Transkei, with no opportunities for cross-border commuter traffic, recorded a substantially higher ratio than Ciskei, where the East London industrial complex and other towns enabled a larger proportion of workers to become commuters rather than migrants.

CASINO STATES

The independence of four of the homelands released them from a host of South African legal restrictions, including those related to gambling. The South African government has retained a highly restrictive attitude to gambling and morally dubious activities. The independent homelands provided the venue for allowing wealthy South Africans to indulge themselves. In this they followed the example set by Botswana, Lesotho and Swaziland after independence in the 1960s, which had encouraged casino development to boost the tourist industry.

The network of casino resorts was designed initially by the Southern Sun Hotel chain of South Africa (Figure 3.25). The result was a spread of enterprises strategically placed with regard to the main centres of the country. The major resort was developed at Sun City in Bophuthatswana, only a relatively short journey from Johannesburg and Pretoria. Because of its situation, it was expanded several times, culminating in the creation of the exotic Lost City. Other casino resorts were established in Bophuthatswana which catered for a wide geographical spread of trade including GaRankuwa, north of Pretoria, to Thaba Nchu near Bloemfontein. The first casino in the Transkei was eccentrically situated on the Wild Coast adjacent to the Natal South Coast, one of the major holiday resort regions of South Africa. Ciskei similarly developed not only the urban casino in Bisho, but a casino resort on the Fish River to attract people from the eastern Cape.

POSTSCRIPT TO STATE PARTITION

State partition has been a most controversial policy wherever it has been pursued in the twentieth century, whether in Ireland, India, Cyprus, or Palestine. The ethnic basis of

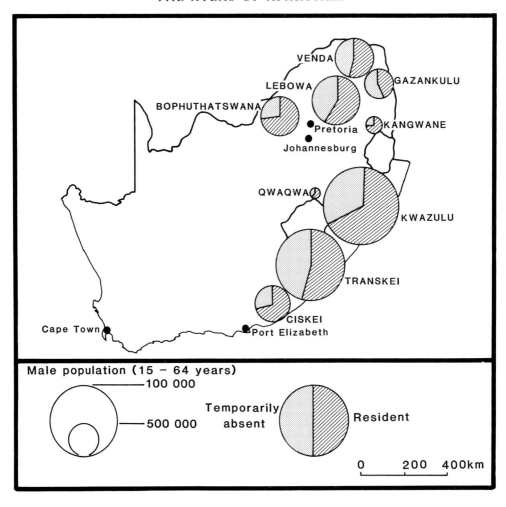

Figure 3.24 Labour migrancy levels, 1970

Source Statistics in T. Malan and P.S. Hattingh (1976) *Black Homelands in South Africa*, Pretoria: Africa Institute

state partition plans have always been regarded as suspect by those groups which lost political powers or land. There has been much argument concerning the basis of the South African government's plans to partition the country. The basic premise that partition was necessary for national survival was not held by the majority of the population, and only by a section, if dominant section, of the ruling White minority for part of the twentieth century. Theoretical academic and political alternatives to the government's partition plans have been suggested periodically to provide a more equitable distribution of the country's land resources. They are worthy of examination if only to expose the essential impracticality of the entire programme.

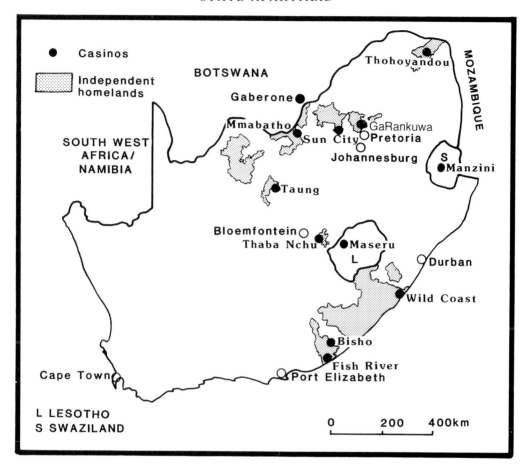

Figure 3.25 Casinos in southern Africa, 1990

Source Information supplied by Sun Hotels

The arguments regarding a more equitable partition were largely academic and dismissed in political circles as being too idealistic and expensive. Thus Blenck and von der Ropp in 1976 put forward a scheme for the division of the country into a Black and non-Black division (Figure 3.26). The basic argument was the acceptance that the Black and White views of the future nature of the South African state were incompatible and that a radical partition acceptable to both sides was worth investigation. The Black–non-Black division appeared as a variant of the indigenous–immigrant dichotomy, assuming that the Coloured and Indian population would identify with the Whites if offered full political rights. The zone indicted thus reflected the historic frontier delimiting the extent of Black settlement, which coincided with regions of poor economic development. Maasdorp later refined the idea by indicating the intricacy of the border zone and suggesting a compromise line. Within the two states accordingly

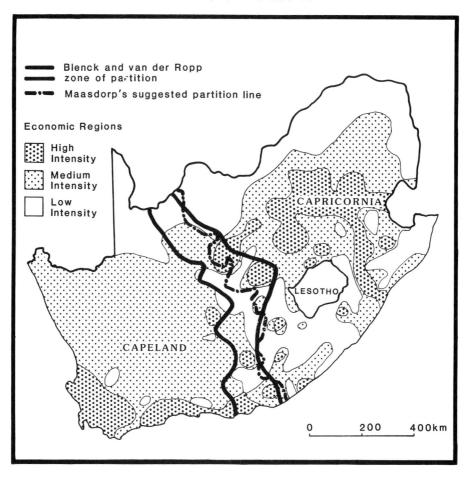

Figure 3.26 Theoretical partition plans

Source Based on J. Blenck and K. von der Ropp (1977) 'Republic of South Africa: is partition a solution?'
South African Journal of African Affairs 7: 21–32, and G. Maasdorp 'Forms of partition',
pp. 107–46 in R.I. Rotberg and J. Barratt (1980) (eds) *Conflict and Compromise in South Africa*, Cape
Town: David Philip

identified, Capeland and Capricorna, the White population would have been in a minority to a Coloured majority in the former and a Black majority in the latter. The basic element of these schemes was the fact that the Witwatersrand and Natal would fall within the Black state, clearly not a viable practical policy for all but a few White political parties.

By the late 1980s the imminent demise of the apartheid era again raised the policy of partition as a political platform. The scale of the rump White state varied from the Conservative Party's attempt to maintain the homeland system with only a few

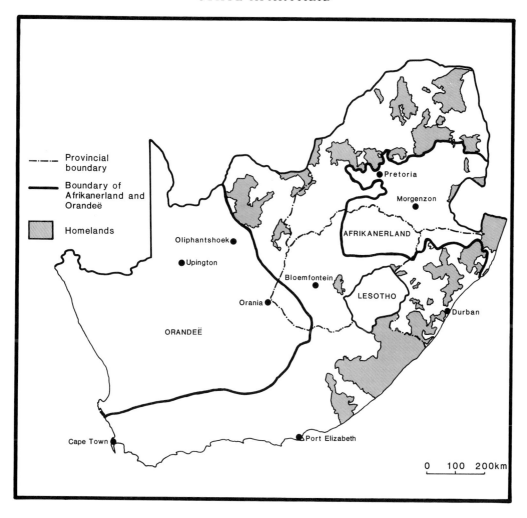

Figure 3.27 Right wing White political partition plans

Source After maps in *Sunday Times*, 19 February 1989, Johannesburg and *Time*, 17 December 1990, New York

modifications, to the Oranjewerkers who initially sought a single village (Morgenzon, in the eastern Transvaal) where they could pursue an all-White labour programme, and so preclude all contact with the surrounding Black population except at official or commercial levels.

Three widely publicized schemes are worthy of note as they represented the attempt to create not a White state but an Afrikaner state. First the Afrikaner Resistance Movement (AWB) propounded the concept of the Boerestaat for the Afrikaner nation. This was narrowly focused on the two republics of the Transvaal (South African

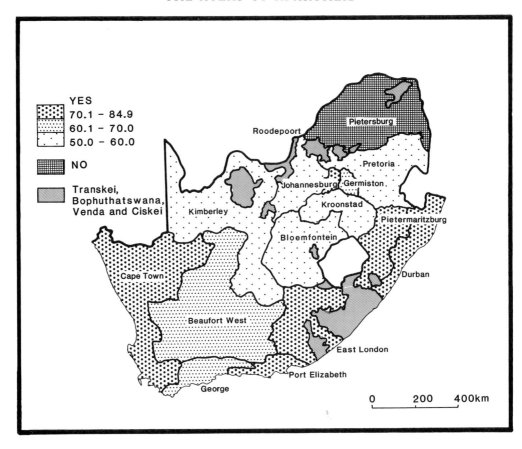

Figure 3.28 Referendum results, March 1992

Source Based on results published in *Eastern Province Herald*, 19 March 1992, Port Elizabeth

Republic) and the Orange Free State which fought against Great Britain between 1899 and 1902 (see Figure 1.6). The area of northern Natal transferred from the Transvaal to Natal at the end of the war remained a territorial claim for such a state, although the interests of the Zulu nation were involved. A variant of this was a more narrowly drawn state excluding the Black homelands and areas of the country such as the central Witwatersrand and Natal which were considered to be heavily Anglicized. The result was the proposal of the Afrikanerland of the Oranjewerkers movement (Figure 3.27). This proposed state retained much of the agricultural heartland of the country as well as much of the industry, and a coastline.

In contrast the state of Orandee proposed by the Afrikaner Freedom Foundation, did not overlap Afrikanerland at all. It was proposed to provide an isolated desert state, which through its very inhospitableness would not be regarded as either desirable or a threat by the Black government which it was assumed would come to power in the remainder of the country. The expounded philosophical basis of the state paralleled that of Israel, in the belief that all Afrikaners would feel more secure if they possessed a homeland in which their culture and language would be preserved by law. It was not intended that more than a small proportion of the Afrikaner nation would reside in Orandee. The town of Orania on the Orange River, abandoned by the Department of Water Affairs, was purchased by the Foundation and attempts were made to put the Afrikaner national development programme into practice.

The real postscript to the partition programme was served by the White electorate in the March 1992 referendum, when over two-thirds of voters accepted the principle of an undivided South Africa. The electorate of only one of the fifteen polling districts wished to pursue the concept of apartheid (Figure 3.28). Remarkably, the northern Transvaal had been included in neither of the partition programmes outlined above because of the overwhelming (97 per cent) Black majority in the region. Perceptively, President F.W. de Klerk remarked that it was fitting that those (the Afrikaners) who had initiated apartheid in 1948 were those who rejected it in 1992.

4

URBAN APARTHEID

Apartheid operated on a local, urban level. The major impetus of the initial years of the National Party administration after 1948 was directed towards the implementation of residential and personal segregation rather than the more philosophically based concept of state partition. Much of the basic legislation affecting Blacks was already on the statute book and was duly enforced with greater ruthlessness. However, the residential segregation of the remainder of the population was a popular issue which enjoyed wide support within the White community of whatever political persuasion. By the 1960s political debate in the country, whether in White or Black communities, was essentially urban based. Urban segregation and the attempts to restrict Black access to the urban areas were thus the dominant political issues of the ensuing decades.

THE POPULATION REGISTRATION ACT

In 1950 the two major pieces of legislation designed to create the new, apartheid city were enacted, namely the Population Registration Act and the Group Areas Act. The Population Registration Act provided for the compulsory classification of the population into discrete racially defined groups. Three basic groups were identified: White, Black, and Coloured. The latter group was split into several subdivisions: Cape Malay, Griqua, Indian, Chinese, and a residual Cape Coloured group. The *ad hoc* and often ambiguous classifications adopted in the colonial and early Union period thus gave way to a rigid system, enforced by statute.

The criteria adopted for classification purposes were based on physical appearance and social acceptability. Thus in the process considerable latitude was theoretically possible, although, in general, the rules were applied to prevent anyone who was not recognizably *and* socially acceptable as a White person from gaining that status. Families were consequently split as members were subjected to such distasteful tests as curliness of hair, skin colour and linguistic ability. The Act also made provision for reclassification where it was considered that descriptions were inaccurate. By the late 1980s as many as 1,000 people a year were seeking reclassification, mostly to become Coloured instead of Black or White instead of Coloured. Indeed between 1983 and

1990 some 7,000 persons officially had their racial description changed (Figure 4.1). In this period only 120 people changed their classification to Black, but 3,561 changed to White. It might also be noted that the children of mixed marriages nearly always took the race of the parent down the political pecking order, thereby keeping the White group as White as possible, although if there was no White parent, the classification of the father was often adopted. Under the Prohibition of Mixed Marriages Act (1949) marriages between Whites and members of other groups were prohibited, while in 1950 the Immorality Amendment Act banned extra-marital relations between Whites and members of other groups. The various prohibitions did not affect marriages and extra-marital relations between members of other groups as only the racial purity of the White group was of official concern.

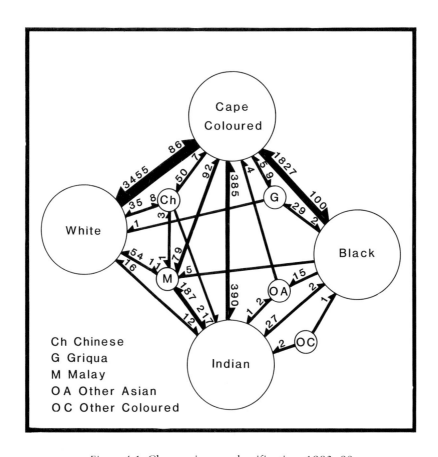

Figure 4.1 Changes in race classification, 1983–90

Source Based on statistics from *Hansard*, 19 June 1991, col. Q2009

THE GROUP AREAS ACT

The Group Areas Act was referred to by the Minister of the Interior, Dr T.E. Donges, as 'one of the major measures designed to preserve White South Africa', when he introduced the bill in parliament.[1] Its conception was to effect the total urban spatial segregation of the various population groups defined under the Population Registration Act. Towns and cities were to be divided into group areas for the exclusive ownership and occupation of a designated group. People not of the prescribed designated group would be forced to leave and take up residence in the group area set aside for their own group. The result was to be total segregation (apartheid), not the piecemeal results of colonial and Union segregationism. Thus social contact between the communities would be reduced to a minimum and competition for urban space legally eliminated.

Administratively, the Land Tenure Advisory Board, established in 1946 to oversee the drawing of Indian areas, assumed the task of drawing up the plans for the new apartheid cities. In 1955 it was renamed the Group Areas Board. The process of declaring group areas was long and complex, involving the formulation of the municipalities' proposals, public enquiries, Board recommendations, and ministerial approval, before the final proclamation was published in the *Government Gazette*. Furthermore, the initiation of a scheme was prompted by the central authorities, often in the face of municipal opposition. However, the reordering of towns and cities was part of the general philosophy of town planning in the post-Second World War era and sustained opposition was rare. Moreover, the administrative machinery for effecting the purposes of the act were not put in place until 1955, when the Group Areas Development Act (after 1966 the Community Development Act) provided a mechanism for expropriation and land development.

Although the supporters of the Act clearly envisaged the inclusion of the Black population within its ambit, the Ministry of Native Affairs and its successors were sufficiently powerful to retain control over the Black population through separate legislation. The Natives Resettlement Act of 1954 provided the mechanisms required to remove Blacks from inner Black areas, while the existing legislation and its modifications enabled the government to control virtually all aspects of Black personal and community life. The practical problems of declaring Black group areas in an era when Blacks could not own land in freehold tenure outside the homelands further confused the issue. The attempt to create a uniform system after the 1950s only resurfaced briefly in the late-1980s. The Group Areas Act for most purposes therefore was only specifically applied to the non-Black population, although the exclusionary clauses did affect Blacks. This most notably applied to the periods of time given to people disqualified under the Act to evacuate or sell properties in proclaimed group areas.

The guidelines for demarcating group area boundaries were drawn up by the Durban Corporation, in the light of its experience in segregating the Indian population before 1950, and were informally adopted by the Land Tenure Advisory Board. The guidelines proposed that group areas be drawn on a sectoral pattern with compact blocks of land for each group, capable of extension outwards as the city grew. Group areas were to be separated by buffer strips of open land at least 30 metres wide, which were to act as barriers to movement and therefore restrict social contact. Accordingly rivers, ridges, industrial areas, railways etc. were incorporated into the town plan. Links between different group areas were to be limited, preferably with no direct roads between the different group areas, but access only to commonly used parts of the city, for example the industrial or central business districts. Originally it was also intended that each group area should be self-sufficient for shopping facilities and that local self-government be introduced. The guidelines were subsequently systematized to form the model apartheid city (Figure 4.2).

In practical terms, given the White dominance in the Board's deliberations, a number of salient points emerge. First, the city centres, with virtually no exceptions, were zoned as part of the White group area. Second, areas zoned for the other groups were highly restricted and peripheral. Third, the sectoral zonings affected not only racially mixed residential areas, but many previously segregated suburbs which fell within the broad divisions of another group.

Armed with these guidelines and a host of suggestions from municipalities, political parties and other interested parties, the Land Tenure Advisory Board began the redrawing of South African cities. If the case of Port Elizabeth is examined, there was a gap of ten years between the drawing up of the first group area proposals (1951) and the first proclamation (1961). In that time the local authorities consulted widely and offered varying schemes, which failed to meet the Land Tenure Advisory Board's requirements of neatness and sectoral contiguity. Those of the Joint Town Planning Committee clearly deviated from the model quite widely (Figure 4.3). The retention of isolated pockets of Coloured or Asian housing which the Municipality wished to leave undisturbed was equally not acceptable to the Board (Figure 4.4).

The final proclamation not only adhered closely to the guidelines but specified stringent evacuation periods for those disqualified, ranging from one to ten years (Figure 4.5). Additional areas were set aside as future group areas, including the racially isolated Black Walmer township. Significantly, the separate Malay area which had been the subject of considerable ingenuity in preliminary planning was abandoned, but the Chinese area survived.

That the 1961 proclamation was not the last word on the subject is attested to by some sixteen subsequent proclamations which extended and modified the original dispensation between 1963 and 1990 (Figure 4.6). The extensive White area was enlarged only slightly but the other group areas required substantial additions. Similar problems were experienced in the other major cities, where the pre-apartheid inheritance was complex.

National progress of proclamation was at first quite slow as few local authorities were able to draw up the necessary plans quickly. By 1960 only 175,000 hectares had been proclaimed. However, in the course of the 1960s a further 580,000 hectares were designated and the majority of the towns and cities in the country were covered by the

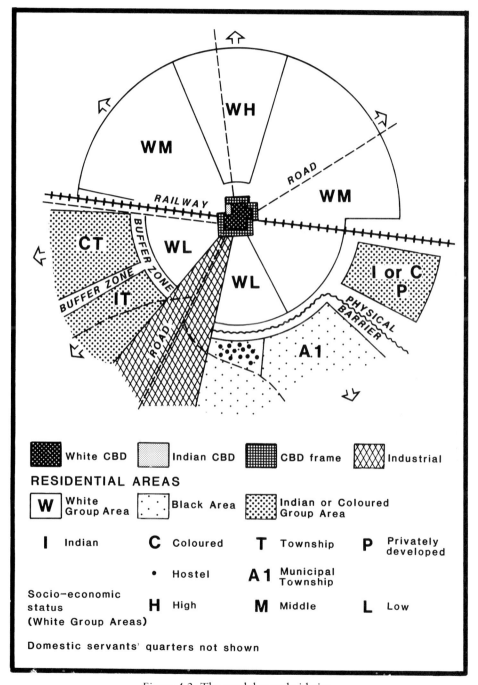

Figure 4.2 The model apartheid city

Source After R.J. Davies (1981) 'The spatial formation of the South African city', *GeoJournal* Supplementary Issue 2, pp. 59–72

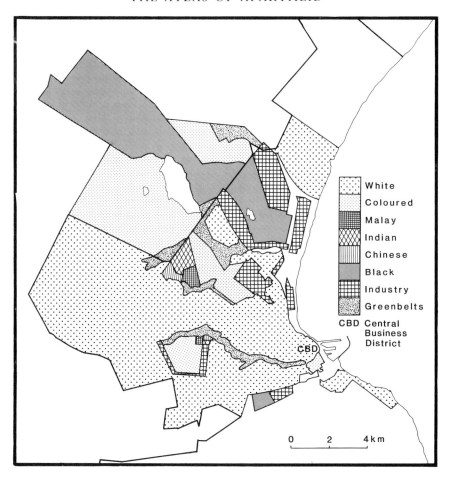

Figure 4.3 Port Elizabeth Joint Planning Committee group area proposals, 1953

Source After W.J. Davies (1971) *Patterns of non-White Population Distribution in Port Elizabeth with Special Reference to the Application of the Group Areas Act*, Port Elizabeth: University of Port Elizabeth

proclamations (Figure 4.7). Thereafter zonings generally were related to extensions for new housing areas and modifications arising from changed plans.

The range of interpretations incorporated into the group area plans was substantial, suggesting that the Board's guidelines were only very broadly applicable. In a number of cities the Group Areas Board was intent upon proclaiming as much land as possible for the White group. The proclamation of an extensive area covering Table Mountain in Cape Town as a White group area in 1958 was evidence of the symbolic use of the Act, as no residential development was envisaged in the area. The most extensive White group area (80,000 hectares) was proclaimed around the administrative capital, Pretoria, where the plans for the whole country were drawn up. This wholesale zoning

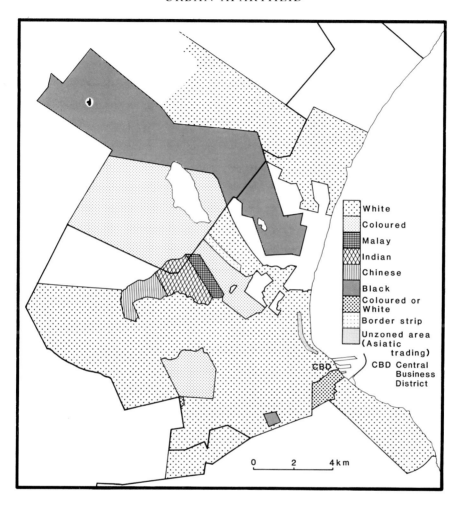

Figure 4.4 Port Elizabeth Municipal Planning Committee group area proposals, 1955

Source After W.J. Davies (1971) *Patterns of non-White Population Distribution in Port Elizabeth with Special Reference to the Application of the Group Areas Act*, Port Elizabeth: University of Port Elizabeth

of towns and cities as White group areas was such that in most cases allowance was made for the needs of the White group for the foreseeable future. In other cases proclamations were restricted to the existing residential and business areas only. In the mining and industrial towns of the Witwatersrand the result was a patchwork of zoned areas and intervening 'controlled' areas, which effectively remained White as a result of the White ownership of the land (Figure 4.8).

In contrast the areas set aside for the other groups were severely prescribed. In general the Indian and Coloured areas were small and peripheral, with few exceptions. The approach adopted by the Board varied from the division of the existing areas of the

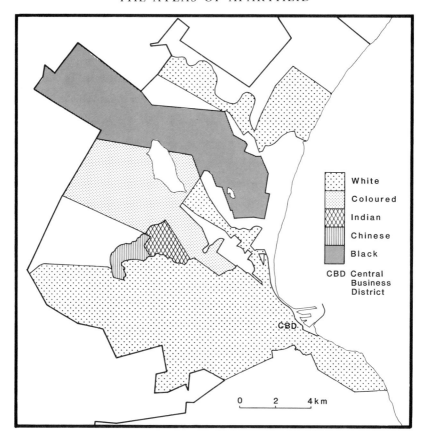

Figure 4.5 Port Elizabeth Group Areas Board recommendations, 1960

Source After W.J. Davies (1971) *Patterns of non-White Population Distribution in Port Elizabeth with Special Reference to the Application of the Group Areas Act*, Port Elizabeth: University of Port Elizabeth

town into zones predominantly occupied by the various groups, to the proclamation of all the existing areas as White and the setting aside of new areas devoid of any building for the other groups. In a number of cases group areas were redrawn in order to exclude Indians from the central business districts.

Although provision was made for the establishment of group areas for the Cape Malay, Chinese, and Griqua populations, few such areas were ever designated. One Cape Malay area, the Malay Quarter of Cape Town, was established, but the majority of the Cape Malay population, even of that city, had to live in the areas designated for the Coloured community. Although several Chinese areas were proclaimed, only one, in Port Elizabeth, was ever developed as the small population was widely scattered in areas such as the Witwatersrand. No Griqua areas were set aside even in East and West Griqualand.

Attempts were made in the Transvaal to centralize the scattered Indian and Coloured

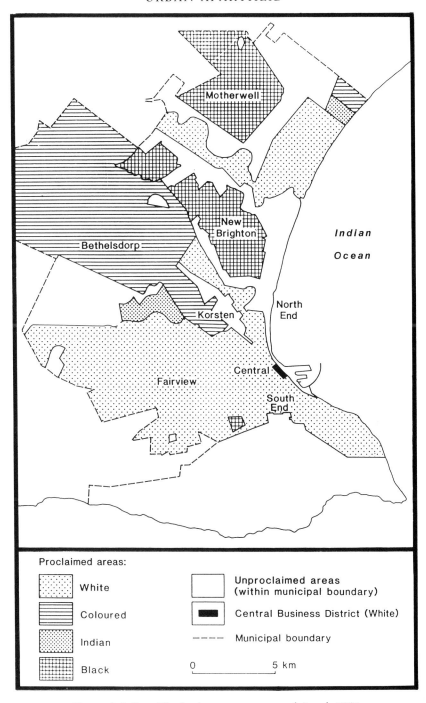

Legend:

Proclaimed areas:

White

Coloured

Indian

Black

Unproclaimed areas
(within municipal boundary)

Central Business District (White)

Municipal boundary

0   5 km

Figure 4.6 Port Elizabeth group areas proclaimed, 1991

Source Based on information extracted from Port Elizabeth municipal maps

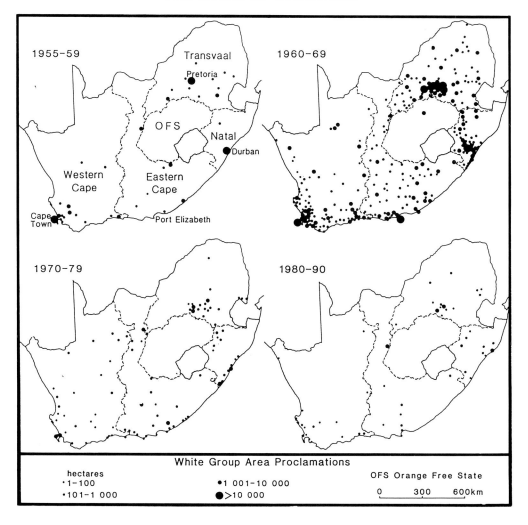

Figure 4.7 Sequence of White group area proclamation

Source Based on information extracted from the files of the Department of Planning, Provincial Affairs and National Housing, Pretoria

urban communities into regional centres. In these centres separate facilities, including schools and clinics, could be provided for communities considered too small to support the basic infrastructure of a group area in each small town. Thus many of the initially designated group areas for the two groups established in the late-1950s and early-1960s were abolished in favour of centralization (Figure 4.9). The same form of centralization was attempted on the East and West Rand until the 1980s where Roodepoort and Boksburg established group areas for the Coloured communities of the two regions respectively, while Krugersdorp and Benoni catered for the Indian community. Only in

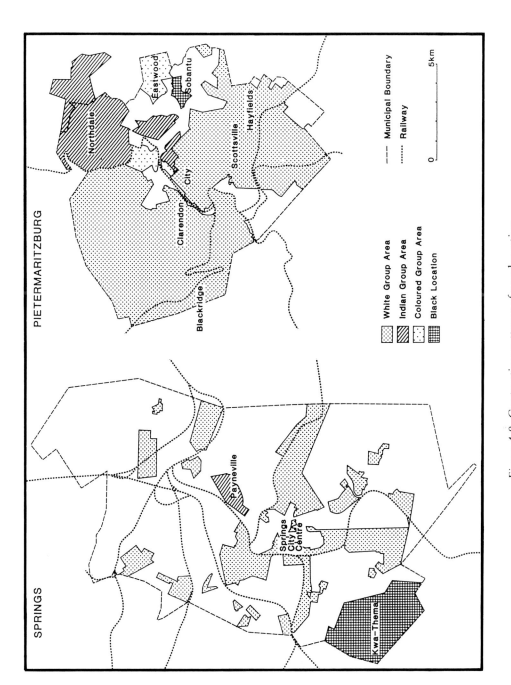

Figure 4.8 Contrasting patterns of proclamation

Source Based on information extracted from the files of the Department of Planning, Provincial Affairs and National Housing, Pretoria

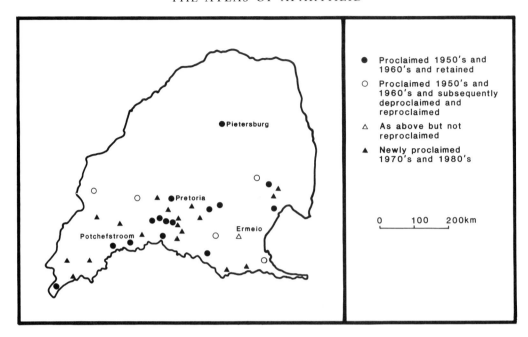

Figure 4.9 Centralization of Coloured areas in the Transvaal

Source Based on information extracted from the files of the Department of Planning, Provincial Affairs and National Housing, Pretoria

the 1970s and 1980s was this policy reversed and the smaller centres and nearly all the Witwatersrand towns provided with separate Indian and Coloured group areas.

Changes in the application of the Group Areas Act took place in the 1980s as constitutional developments necessitated a more generous approach to problems of the Indian and Coloured communities. Thus there was a decline in the extent of White proclamations, indeed in the course of the 1980s more White areas were being deproclaimed than new areas proclaimed. However, there were notable extensions to the Indian and Coloured areas allowing for expansion. In some cases the extended areas came from deproclaimed White areas. In contrast the Chinese were incorporated administratively into the White group in 1984 necessitating the inclusion of the Chinese group area into the White area. It should be noted that honorary White status did not include the franchise for the community.

Most significantly a number of inner suburbs, initially proclaimed White, were reproclaimed for the communities originally resident in them. The most notable suburb was Woodstock in Cape Town, reproclaimed for the Coloured population (Figure 4.10). Mayfair in Johannesburg was similarly treated for the Indian community. Thus the grand apartheid pattern of sectoral racial divisions of the city as originally conceived was interrupted by a number of non-conforming enclaves in the early-1980s.

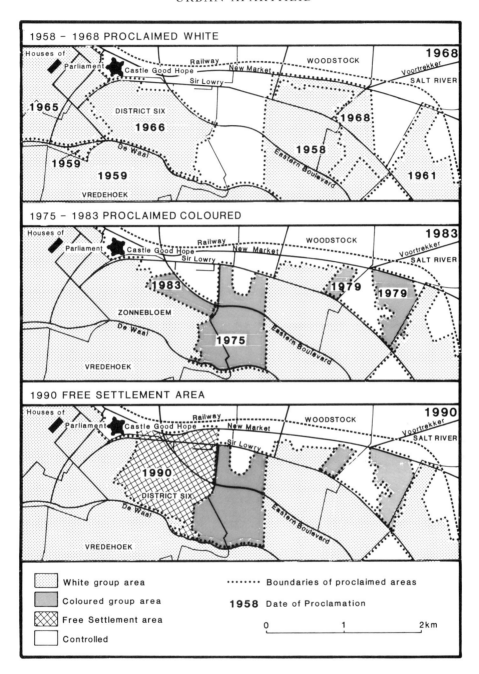

Figure 4.10 Changed inner areas zonings, Cape Town

Source Based on maps supplied by the Cape Town municipality

POPULATION RESETTLEMENT

The administration of the Group Areas Act was limited until the later 1950s by the lack of administrative procedures for converting official intention into reality. However, in 1955 the Group Areas Development Board, later the Community Development Board was given the powers to purchase, sell, and develop land, with extensive powers of expropriation. Accordingly in the late-1950s and the 1960s surveys of disqualified people and affected properties were undertaken, to assess the extent of property expropriation and probable population movement involved in the carrying out of the schemes to fit the population to the newly drawn group areas boundaries (Figures 4.11 and 4.12). The detailed house by house survey of mixed areas provides one of the best indicators of the extent of the racial integration which still survived in the late-1950s and early-1960s.

It is not possible to establish accurate figures for the number of people displaced under the Group Areas Act. The use of the Slums Act and other measures to attain the ends of the Group Areas Act, makes official statistics appear much smaller than in reality. However, the conservative figures produced by the authorities suggest that in the period up to 1984, when the administration of the Act underwent a major reorganization, a total of 126,000 families were displaced under the Group Areas Act. Of these only 2 per cent were White, such was the skill of the Group Areas Board in drawing the boundaries to interfere with the White community as little as possible. By contrast, in many towns, virtually the entire Indian and Coloured populations were displaced. This was particularly evident where scattered pre-1950 municipal housing estates were rezoned for White occupation.

The reorganization of the population distribution patterns was remarkable. After 1950 most municipalities were forced, through central government financial sanctions, to operate as though effective group areas had been proclaimed, so far as the provision of new state housing was concerned. Newcomers migrating to the towns were directed to racially homogeneous suburbs, while new families seeking their own homes for the first time were similarly directed. Forced migration from the central suburbs took place in the 1960s and early-1970s, parallel with expropriation and demolition.

The transformation of the patterns of population distribution is thus striking. In central Cape Town the contrast between the distribution recorded in the 1951 and 1985 censuses at census tract level, in consequence of the demolition of District Six and the expulsion of many other people as a result of the Group Areas Act, gives some indication of the detailed workings of the system (Figure 4.13). At a city-wide scale of suburbs, the transformation of the distribution of the population of Port Elizabeth between 1960 and 1991 indicates the extent of resettlement and the essentially peripheral nature of the new residential areas for those people not classified as White (Figures 4.14 and 4.15).

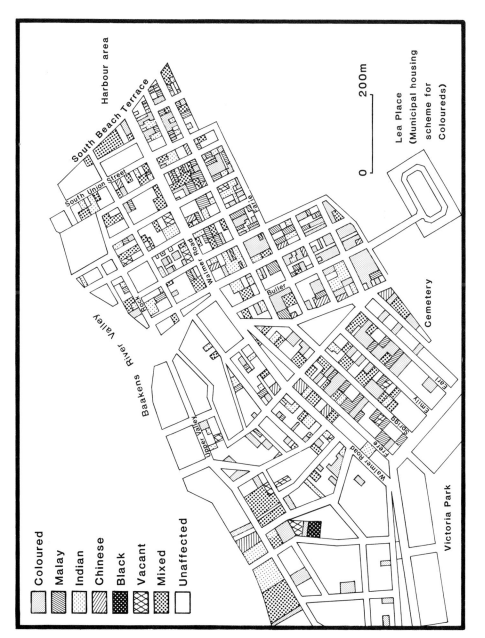

Legend:
- Coloured
- Malay
- Indian
- Chinese
- Black
- Vacant
- Mixed
- Unaffected

South Beach Terrace

Harbour area

South Union Street

Baakens River Valley

Walmer Road

Rock

Buller

Upper Valley Road

Walmer Road

Victoria Park

Cemetery

Frere

Emily

Spring

Earl

Lea Place
(Municipal housing
scheme for
Coloureds)

0   200m

Figure 4.11 Disqualified persons in South End

Source After J.G. Nel (1988) *Die Geografiese Impak van die Wet op Groepsgebiede en Verwante Wetgewing op Port Elizabeth*, Port Elizabeth: University of Port Elizabeth

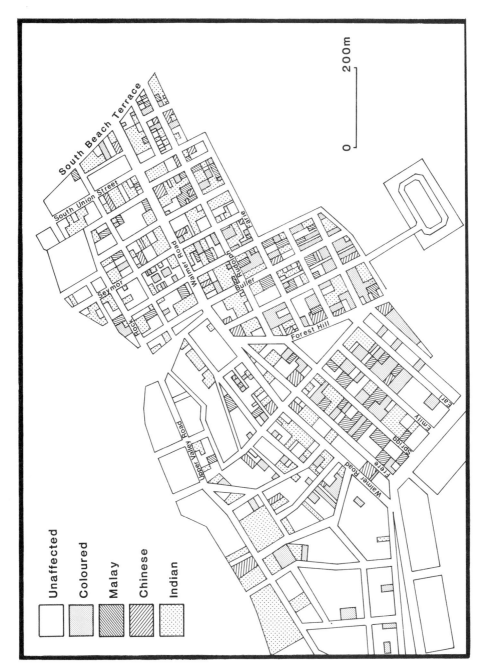

Figure 4.12 Affected properties in South End

Source After J.G. Nel (1988) *Die Geografiese Impak van die Wet op Groepsgebiede en Verwante Wetgewing op Port Elizabeth*, Port Elizabeth: University of Port Elizabeth

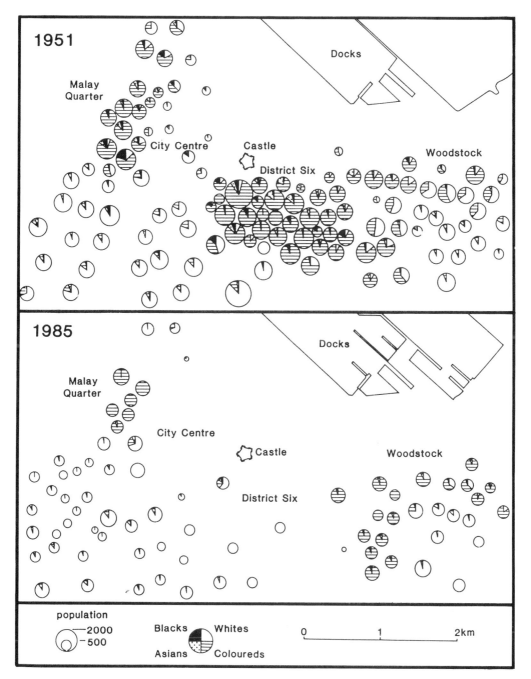

Figure 4.13 Distribution of population in central Cape Town, 1951 and 1985

Source Based on information extracted from the files of the central Statistical Services, Pretoria

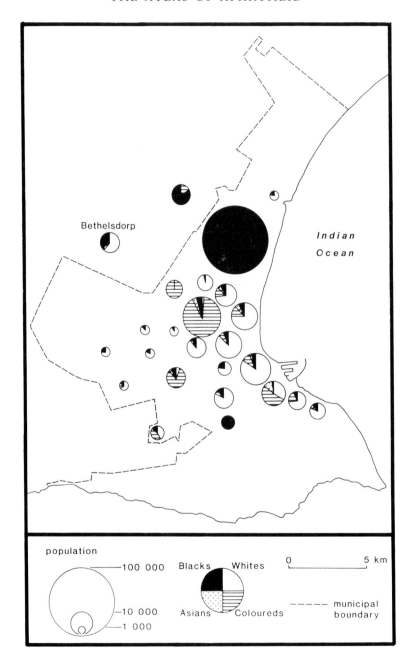

Figure 4.14 Distribution of population in Port Elizabeth, 1960

Source Based on information extracted from the files of the Central Statistical Services, Pretoria

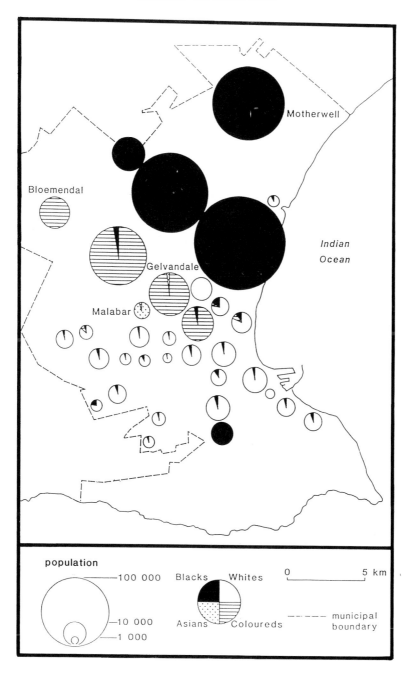

Figure 4.15 Distribution of population in Port Elizabeth, 1990

Source Based on information extracted from the files of the Central Statistical Services, Pretoria

BLACK RESETTLEMENT

The removal of the Black population living in the central or White designated urban areas was addressed in the 1950s under the battery of measures passed by previous administrations and through the tightening of residential control legislation through the euphemistically named Abolition of Passes Act (1951). The Natives Resettlement Act of 1954 provided for the removal of those owners and tenants with legal rights in urban freehold areas (Figure 4.16). Inner Black owned suburbs or individual plots were expropriated and the population resettled on the urban periphery or, where possible, in the homelands. The destruction of Sophiatown in Johannesburg is the most notorious of these removals. Inner Black locations were also demolished or transferred to either Indian or Coloured occupation. A few, such as Wattville at Benoni, survived although threatened with destruction for over forty years. The result of these policies was a substantial movement of population numbering approximately 750,000 people to new Black townships on the periphery of the cities.

New legislation, the Natives (Urban Areas) Amendment Act of 1955, was introduced to remove such concentrations of Blacks as the servants living in central city blocks of flats. Dr Verwoerd suggested that the arbitrary number of five Black servants per block of flats was a 'wise' number to prevent overcrowding in the growing flatlands of Johannesburg and other cities. The number involved in this particular movement in central Johannesburg was approximately 10,000, most of whom were rehoused in single sex hostels in Soweto.

The majority of the Black townships were inherited from the previous era. The Natives (Urban Areas) Consolidation Act of 1945 with its various amendments provided the basic framework for Black township control until the implementation of the Black Communities Development Act of 1984. It is worth noting that the area set aside for Black townships until the late-1980s was small. This reflected the government's intention that the Black population was, with few exceptions, temporary. The development of the Black states, furthermore, was ostensibly to provide the attraction for Blacks to live there and so depart from the White areas. This had the implications that only minimal housing was provided and private building was discouraged, in favour of homeland development. Indeed Dr Verwoerd predicted that by the 1970s Black migration to the cities would be reversed as the homelands would provide a greater attraction for workseekers. Government policies were pursued with this unrealistic goal in mind.

However, the movement of people from the rural areas, and the quest for more urban living space remained a constant theme of the period after 1948. In 1955 the Eiselen line was drawn around the western Cape, and subsequently extended, in order to declare the region a Coloured Labour Preference Area (Figure 4.17). Under this policy only Coloured labour was to be employed in the region, unless it could be shown that there were no suitable workers available. The Coloured rural reserves, notably in the arid district of Namaqualand, were occasionally referred to as the basis of a

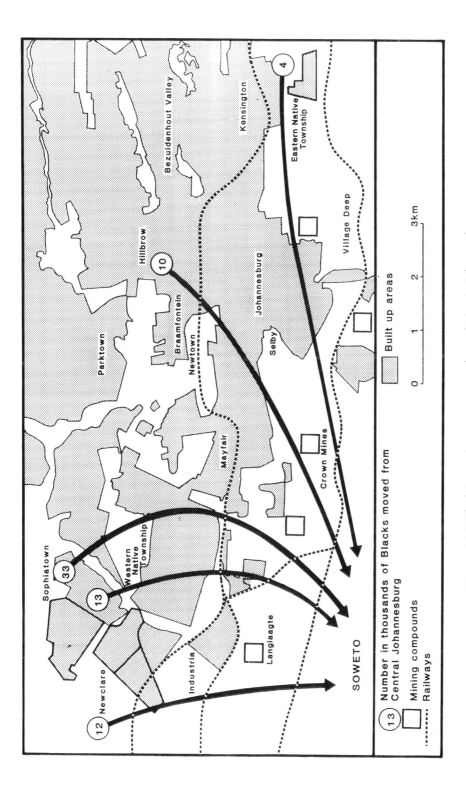

Figure 4.16 Black inner city movements of population in Johannesburg

Source Based on 1951 and 1960 census enumeration sub-district information extracted from the records of the Central Statistical Services, Pretoria

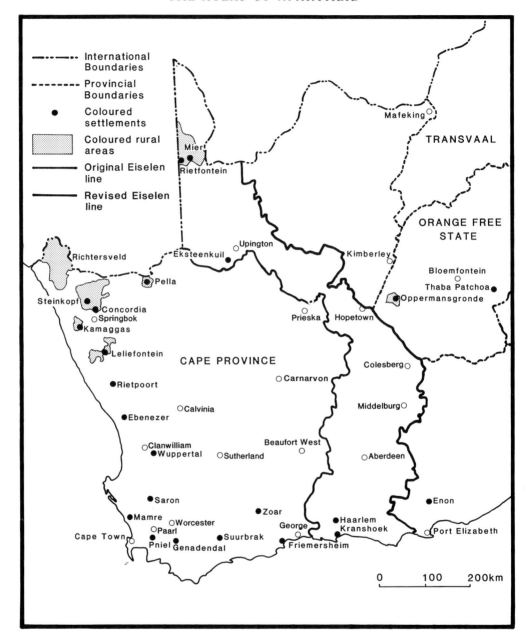

Figure 4.17 Eiselen Line and Coloured settlements

Source Based on A. Lemon (1976) *Apartheid: A Geography of Separation*, Farnborough: Saxon House;
G. Maasdorp, 'forms of Partition', pp. 107–46 in R.I. Rotberg and J. Barratt (1980) *Conflict and
Compromise in South Africa*, Cape Town: David Philip, and information supplied by the Provincial
Adminstration of the Orange Free State

Coloured homeland, but no such entity ever entered official policy.

Accordingly the official intention behind the Eiselen line was to remove virtually all Blacks from the region. This policy was again a failure but was only recognized as such in 1983. The policy resulted in the emergence of a serious Black housing crisis in the Western Cape as the construction of state housing for the community ceased. Consequently Black migrants were forced to squat illegally in settlements such as Crossroads on the periphery of Cape Town. The official response was to demolish the shacks and send the people involved back to the homelands. In 1980 alone some 16,000 people were arrested for contravening the strict control regulations.

It was only with the lifting of influx control in 1986 that a massive reversal of this restrictive policy became evident. The provincial authorities then responsible for the designation of Black Development Areas sought out land on a substantial scale. In 1990 alone some 17,000 hectares were designated as Black Development Areas, compared with 124,000 hectares in the previous ninety years. During the final five months of zoning a further 6,000 hectares were designated, including six areas on the last working day before the repeal of the racial zoning legislation. No better illustration of the perseverance of the bureaucratic machinery could be presented. Black Development Areas remained remarkably crowded with densities on average more than twice that of the Coloured or Asian group areas, and eight times that of the White group areas.

BLACK ETHNIC SEGREGATION

The policy of state partition intruded into the planning of the apartheid city as the Black population was not officially regarded as one entity. Ethno-linguistic differences noted for the purposes of establishing separate nation-states were emphasized in order to link the urban Black population to their homelands, where they were to obtain political rights. In 1954 the government instituted the policy of ethno-linguistic segregation to 'assure the simplified and improved education of children in house [sic] languages; to maintain tribal discipline; to assist the efficient functioning of Bantu [Black] authorities; to simplify municipal control; and to make for more harmonious living among the Bantu'.[2]

Accordingly the Black areas of towns were zoned for the different linguistic groups. In the majority of cities only one such group was present in any numbers. Thus the Black suburbs of Durban were over 90 per cent Zulu-speaking, while the Black population of Port Elizabeth was 97 per cent Xhosa-speaking. In such circumstances there was little to be achieved in segregation between different linguistic groups. However, on the Witwatersrand the Black population had been drawn from all over South Africa, and no such dominance by one group was evident (Figure 4.18). In 1985 some 26 per cent of the population of the Witwatersrand Black townships was Zulu-speaking and 15 per cent Tswana-speaking. The Xhosa, North Sotho and South Sotho also each exceeded 10 per cent of the total Witwatersrand townships population, although local concentrations varied substantially.

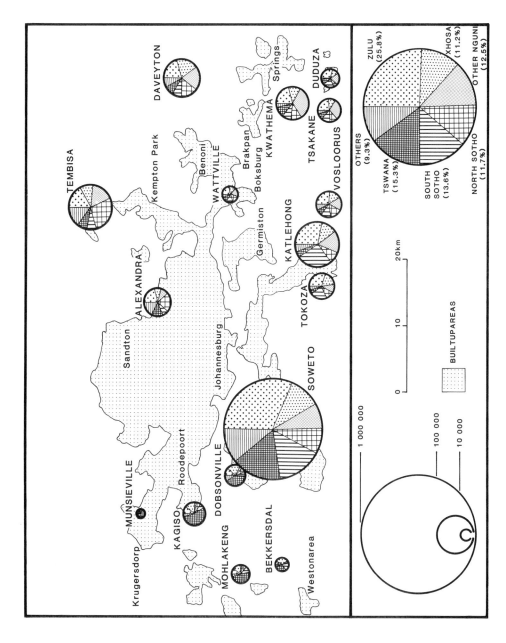

Figure 4.18 Ethnic composition of Witwatersrand Black townships, 1985

Source Based on information extracted from the files of the Central Statistical Services, Pretoria

Segregation was achieved through the manipulation of housing allocation systems for new arrivals in the Black townships. The policy was implemented in varying ways and without the determination evident in the enforcement of group area policies. In the older inner townships no zoning was possible as housing was allocated on an 'ethnic-blind' manner according to the availability of housing and new residents. In expanding townships, the new extensions alone could be zoned, leaving some areas ethnically integrated and others segregated. Only in the completely new townships planned and settled after 1954 could the policy be enforced with some degree of completeness. Owing to the magnitude of the housing programme and the speed at which resettlement from the remainder of the cities took place, housing managers were more usually concerned with supplying houses to those in need, than observing the ethno-linguistic segregation policies of the government. Thus the majority of townships were zoned on a simpler three-fold Nguni, Sotho and 'Other' basis. Only at Daveyton was an attempt made to segregate each group from virtually every other (Figure 4.19).

As a measure of the 'success' of the policy it should be noted than in the older townships which experienced no expansion after 1954, ethnic segregation levels were remarkably low, usually signifying little more than random distributions. In most of the new and expanded townships segregation between the major grouping was substantially higher, suggesting a degree of success in implementing the policy. However, only in Daveyton did groups such as the Xhosa and Zulu experience segregation, elsewhere they were integrated with apparently random distributions relative to one another. Nowhere were intra-Black indices of segregation as high as those between the White and Black populations.

RESIDENTIAL SEGREGATION LEVELS

One of the basic questions confronting an assessment of urban apartheid is the degree to which the official plan was converted into reality. In examining the levels of residential segregation attained by apartheid social engineering the populations enumerated at tract or sub-district level at the time of the 1991 census have been analysed. The results of the 1991 census have serious problems attached to them due to survey sample methodology and the policy of non-co-operation adopted by the African National Congress during a period of political strife. Furthermore, by 1991 the results of the easing of several apartheid laws were beginning to be noticeable in White group areas. The census of 1991, despite its imperfections, may be regarded as a measure of the apartheid extreme.

The marked regional variations in segregation levels noted in the 1951 census remained in evidence, with a few exceptions, but all indices were raised to remarkably high levels. The full impact of apartheid planning was experienced in the Western Cape, which had exhibited the greatest degree of inter-racial residential integration before the implementation of the Group Areas Act. By 1991 it had become one of the most highly segregated regions of the country. The patterns which emerged were those of almost

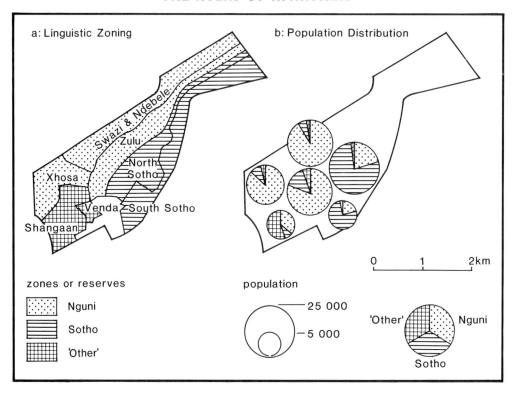

Figure 4.19 Ethnic zoning in Daveyton

Source Based on South Africa (1954) *Planning of Residential Areas for Bantu in Urban Areas: ethnic grouping*, Pretoria: Department of Native Affairs and 1985 census enumerators' returns extracted from the Central Statistical Services, Pretoria

total segregation. Only 8.6 per cent of the urban population of South Africa (excluding the homelands) lived outside their designated areas by 1991. Significantly only 0.2 per cent of the White population lived outside the White group areas, whereas 10.5 per cent of the Coloured population, 12.0 of the Asian and 12.8 per cent of the Black population did so (the latter figure would be reduced to under 10 per cent if the urban population of the homelands is considered). Most of these persons lived in segregated areas zoned for another group, notably Blacks in old central locations due for demolition or in barrack accommodation adjacent to industrial sites. Others not conforming to the pattern included the declining numbers of domestic servants living on the properties of their employers. However, despite forty-one years of official attempts to reduce the numbers of Black domestic servants, few census tracts in White towns failed to record at least 5 per cent of the population as belonging to other groups.

Marked variations, nevertheless, occurred within South Africa. The White popu-lations of the towns of the Orange Free State and the Eastern Cape were significantly more segregated than those in Natal and the Transvaal. This state of affairs parallels the colonial and early Union pattern closely, albeit at much higher levels of separation. Remarkably, there was no significant difference in segregation levels between large and small towns (Figure 4.20).

The Coloured population had undergone an exceptional transformation as a result of apartheid urban planning. Both the Western and Eastern Cape were significantly more segregated than either Natal or the Orange Free State by 1991 (Figure 4.21). Again the nineteenth-century heritage is evident in that the Orange Free State system of housing the Coloured population in the Black locations was only undergoing slow change by 1991, while the Natal Coloured community had traditionally been more

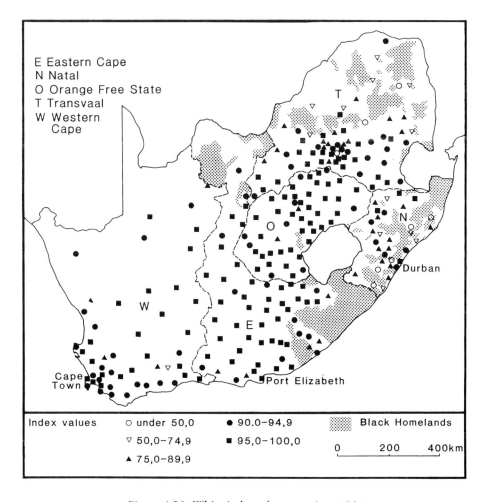

Figure 4.20 White index of segregation, 1991

Source Indices calculated from the census enumerators' returns of the Central Statistical Services, Pretoria

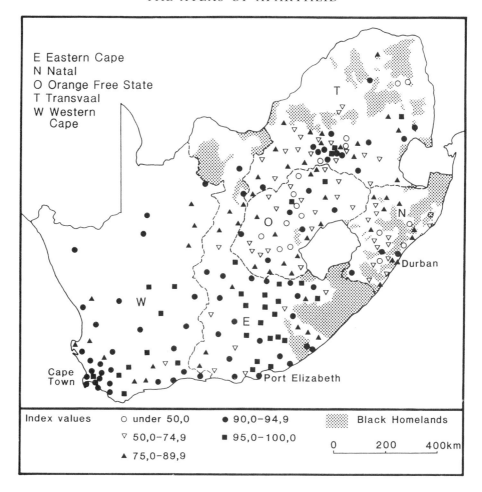

Figure 4.21 Coloured index of segregation, 1991

Source Indices calculated from the census enumerators' returns of the Central Statistical Services, Pretoria

closely connected with the White population. However, the dramatic transformation of the Cape towns, notably those in the Western Cape, is most evident.

The Asian population had been subjected to close regulation since initial immigration in the nineteenth century. Thus segregation levels were high, but with significant regional variations, reflecting the colonial patterns (Figure 4.22). The absence of Indian communities in the Orange Free State is immediately apparent. Elsewhere the Indians in the Transvaal were significantly more segregated while those in the Western Cape remained less segregated. In the latter case the link through religion between the Cape Malays and the Indian Moslems is notable, often making clear-cut classifications impossible.

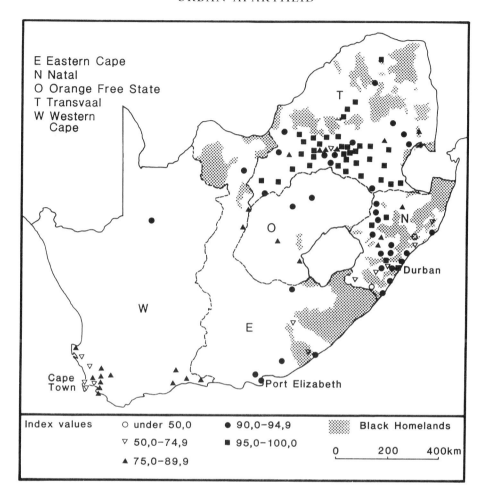

Figure 4.22 Asian index of segregation, 1991

Source Indices calculated from the census enumerators' returns of the Central Statistical Services, Pretoria

The pattern of Black segregation is more complex as the effects of the incorporation of Black communities into the homelands need to be considered (Figure 4.23). Accordingly in Natal and parts of the Transvaal and eastern Cape Province the Black suburbs of a number of towns were included in the neighbouring Black state. The most notable examples are in Durban and East London where virtually the entire Black suburban areas have been built in the homelands or were subsequently transferred to them. In other cases the Black population was rehoused at considerable distances from the original town and no physical connection was evident. The construction of Botshabelo some 30 kilometres from Bloemfontein is a particularly striking example, as some 500,000 people lived in the settlement by the late-1980s. Similar, if smaller, all-

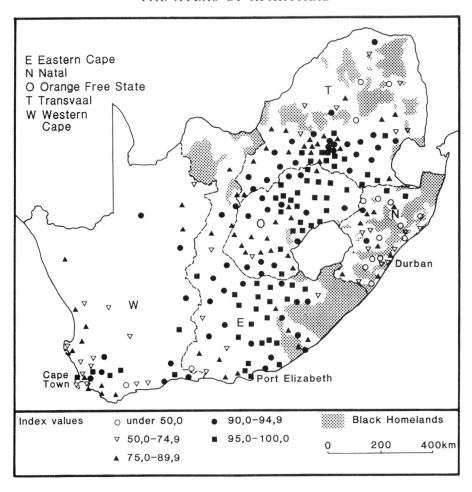

Figure 4.23 Black index of segregation, 1991

Source Induces calculated from the census enumerators' returns of the Central Statistical Services, Pretoria

Black towns were built elsewhere in the homelands to serve a distant White town. Thus the significantly lower segregation levels particularly noted in Natal towns must be regarded with caution. However, the significantly higher levels recorded in the Orange Free State again reflect the colonial pattern.

The administrators of apartheid planning consequently had achieved virtually total segregation in residential patterns in most South African cities by the mid-1980s. Indeed, segregation levels by the time of the 1970 census indicate that in the majority of cities implementation of segregation was nearly complete. This suggests that by the early 1990s remarkably few urban dwellers had lived in racially and ethnically integrated conditions and the current process of dismantling apartheid offers unknown challenges to those involved. In this respect apartheid urban planners may be regarded as having succeeded in their aims by the early-1980s. However, pressures in the 1980s led to modifications away from this rigid model.

BUSINESS AREAS

The Group Areas Act was concerned not only with residence but also with business activities. Thus in continuation with the flow of pre-1948 anti-Indian trading measures, one of the prime purposes of the Act was to remove Indian businessmen from White areas. The Group Areas Act specified disqualification from carrying on a business except in the group area of a member's racial classification. Accordingly in the various central and other business districts trading was restricted to White people. In Natal the dual structure of central business districts was directly affected. In the majority of Transvaal towns a similar structure existed, although the Asiatic bazaars had been designated earlier. Elsewhere Indian traders occupied premises among other groups, although rarely in sections of the financial sector of the central business district.

The expropriation of Asian- and Coloured-owned businesses proceeded with the declaration of group areas. It was possible for shop owners to gain temporary exemption from the terms of the Act through the permit or appeal systems. However, the numbers surviving orders to cease business declined, and only in the late 1970s did Chinese traders begin to obtain permanent exemption from the Act in certain sections of the major cities. The principal area of survival was a section of the Indian central business district in Durban. The group areas proclamation disqualified the residents from living there but permitted trade to be continued (Figure 4.24). This was an unsatisfactory arrangement as the traders usually lived above the business premises. Furthermore, only a restricted part of the Indian trading area was allowed to survive, and not the fringe or adjacent residential areas. However, in 1984 the remaining residents were permitted to stay.

The apartheid ideal of traders being restricted to their respective group areas was clearly impossible for communities such as the Indians who depended disproportionately upon trade for a livelihood. Suitable business locations were essential for traders who relied upon servicing the entire urban population. This was recognized in 1957 with the concept of Free Trade Areas in which Whites, Coloureds and Asians could buy property and conduct business. Few such areas were ever developed. In Port Elizabeth the small Indian businesses displaced from other parts of the city were allowed to relocate on the edge of the industrial area. In Johannesburg the Indian businesses removed from the centrally situated Diagonal Street area were offered sites in a new Oriental Bazaar to the west of the central business district. Similar Oriental Bazaars were established in other cities.

In 1986 as part of the government's attempt to co-opt the Indian community into the new constitutional dispensation, restrictions on trading were relaxed. Free Trading Areas were proclaimed in the majority of central business districts. In the main the areas

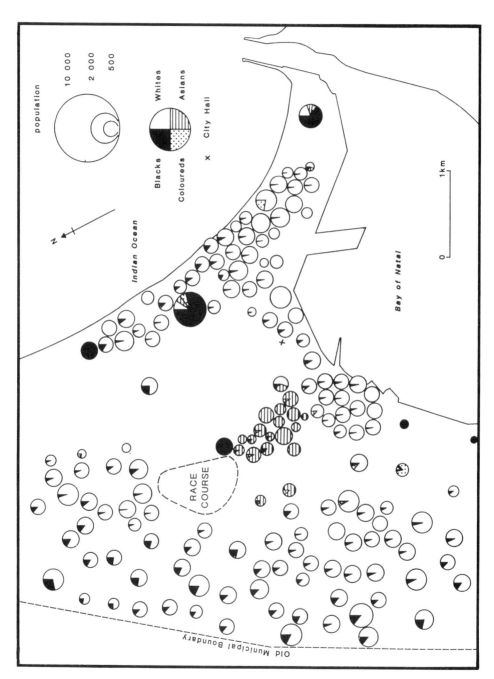

Figure 4.24 Distribution of population in central Durban, 1985

Source Based on information extracted from the files of the Central Statistical Services, Pretoria

involved were small, coinciding with the effective central business district, but excluding small suburban shopping areas, which remained almost exclusively in White hands (Figure 4.25). However, the application of the regulations became progressively more flexible with the result that through the permit system other areas were opened to traders of other groups, prior to the repeal of all regulations in 1991.

It should be noted that although the authorities were anxious to remove Indian and other traders from White areas, there was little concern with other group areas, where Indian, Chinese and Coloured traders were allowed to continue with little interference. Black areas were excluded as only Black owned businesses were permitted within the designated Black areas. These were restricted as limits were placed upon the size of stores (initially 150 and then 350 square metres) and the range of goods (only daily convenience goods) that could be sold until the late-1970s. Only one licence could be

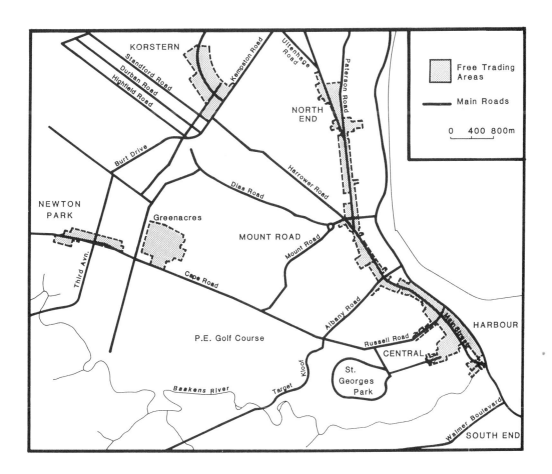

Figure 4.25 Free Trading Areas, Port Elizabeth, 1986

Source Based on maps supplied by the Port Elizabeth municipality issued to an individual and policy decisions decreed that successful businessmen be directed to the homelands. Thus few shopping areas in Black suburbs developed to any size before the 1980s. Alcohol sales, for example, long remained a state monopoly in Black areas, as a dubious source of revenue for the municipalities and subsequent Administration Boards. It had been a principle of the administration of the Black townships since colonial times that no expense should fall upon the White ratepayers. The municipal and later state monopoly of alcohol sales provided the minimal revenues required to administer the townships.

FREE SETTLEMENT AREAS

A further relaxation of the original apartheid plan came into effect in 1990 as a result of the Free Settlement Areas Act passed two years earlier. The workings of the Group Areas machinery were adversely affected by the decision of the Transvaal Supreme Court in the Govender case in 1982, when it was ruled that evictions under the Group Areas Act could only be effected if alternative accommodation could be offered. In view of the massive housing shortage in the Coloured, Indian and Black residential areas, alternative accommodation was rarely available. Thus evictions under the Act virtually ceased. No amending legislation was introduced, as the government at the time wished to attract the Coloured and Indian communities into new constitutional structures.

The result was an influx of Coloured, Indian and Black people into the inner suburbs and a few other areas of the major cities. The high density flatlands in particular attracted people seeking accessible and often single accommodation near the city centre and also wishing to flee from overcrowded housing in the peripheral townships (Figure 4.26). Other inner suburbs where some members of either the Indian or Coloured communities had been able to retain their properties also attracted returnees. Hence Mayfair in Johannesburg attracted Indian families, while elsewhere inner suburbs attracted a wide range of persons, including European immigrants. In this manner 'grey areas' were brought into being throughout the major centres of the country, reproducing, on a highly limited scale, the inner mixed suburbs of colonial times.

The government sought to control and limit the extent of this development and so passed the Free Settlement Areas Act of 1988. The Act introduced the concept of areas where people of all population groups could live. A wide range of proposals, both official and unofficial, were put forward to implement the purposes of the Act. In a number of cases City Councils proposed that entire cities be opened to settlement by all, in others small new suburbs were offered, as though 'grey' represented another 'colour' on the group areas map, necessitating the establishment of separate local administrations! The majority of the areas approved were for such undeveloped peripheral zones, where no one would be forced to live in an integrated society. The final area proclaimed at Uitenhage in 1991 was of this nature (Figure 4.27). Included in this category was the proclamation of the demolished section of District Six in Cape Town as a Free Settlement Area (see Figure 4.10). Few settled areas were proclaimed as the

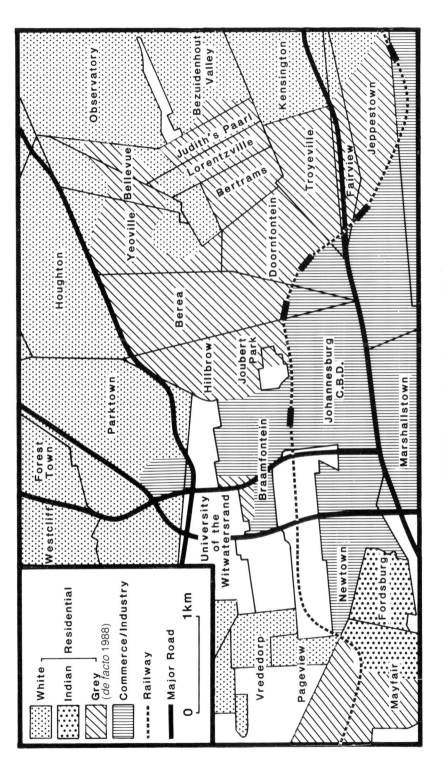

Figure 4.26 Grey areas in Johannesburg, 1988

Source After S.P. Rule (1989) 'The emergence of a racially mixed residential suburb in Johannesburg: the demise of the apartheid city?' *Geographical Journal* 155: 196–203

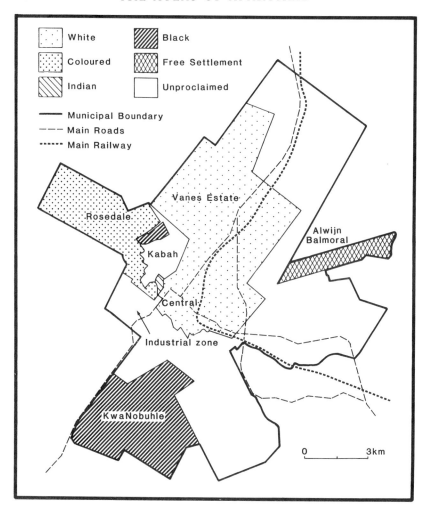

Figure 4.27 Free Settlement Area in Uitenhage

Source Based on maps supplied by the Uitenhage municipality

majority of enquiries into the objections were not completed when the Free Settlement Areas Board effectively ceased to function in February 1991. Only fourteen of the fifty-six formal applications for the establishment of Free Settlement Areas were taken through to proclamation.

NEW TOWNSCAPES

The apartheid city was deliberately created by the South African government over a period of forty years. The broad zoning and governmental control over the detailed

138

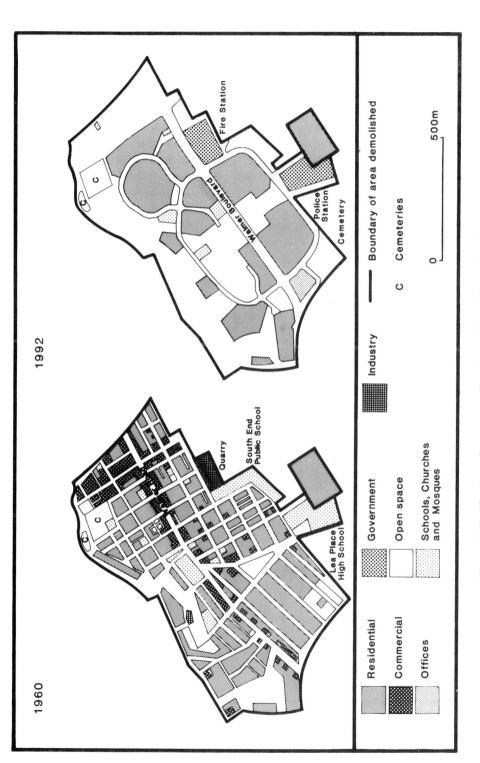

Figure 4.28 The Transformation of South End, Port Elizabeth, 1960–92

Source Based on information from *Donaldson's Port Elizabeth Directory 1959/60, 1959*, Port Elizabeth and air photographs for 1960, and fieldwork, 1992

planning of the city is evident in a variety of aspects. Extensive new residential areas established by the government for the Coloured, Indian and Black populations were built with houses of remarkable uniformity, which were only softened as tenants and owners began the process of house improvement, through new building extension, repainting and gardening. Nevertheless, many housing estates remained much as they were initially in appearance, as the inhabitants were too poor or impermanent to afford to improve the properties. The appearance of uniformity, particularly in the Black areas, was enhanced by the speed with which the areas were built. Half the formal housing of Soweto was constructed in a period of under twenty years from the early-1950s. Only in the 1980s were private developers active in the Black areas, providing a range of housing types and environments. Even so the street plans and plot sizes were still controlled by the authorities. In contrast the White extensions to the cities have followed the free market principles of American cities, with the extensive sprawl of smallholdings, unused farmland and patchy suburban development.

Within the inner core the apartheid city again followed the twin aspects of the First World, capitalist, modern high-rise business districts, with surrounding zones of transition. However, strong evidence of Second World, socialist planning is present in most cities where inner, integrated areas were demolished, providing the authorities with a reserve of land to be used for governmental purposes. The process predates the official adoption of apartheid in 1948. The civic centre at Kimberley was built on the site of the Malay Camp, which was demolished as a clearance scheme under the Slums Act. Expropriated land thus provided room for the construction of new government buildings close to city centres. In Cape Town and Durban the new technicons were built on the extensive areas of demolished houses in District Six and Warwick Avenue respectively, while in Port Elizabeth a new police station and the main fire station were built in South End as part of the radical replanning of the suburb (Figure 4.28).

In a number of cases the land was subsequently used for White housing, resulting in the appearance of modern inner suburbs close to the city centre. The transformation of the predominantly Black suburb of Sophiatown into the White suburb of Triomf is a particularly emotive example, as the renaming suggests. In other cities the cleared expropriated land was left derelict as no decision could be taken on its future which did not involve political controversy. The largest portion of District Six in Cape Town falls into this category as subsequent residential redevelopment was considered inappropriate and the derelict area served as a political reminder of opposition to government policy. Elsewhere houses which were not demolished were resold to Whites as part of a 'gentrification' programme.

The enforced expropriation of property by the government for political purposes poses a substantial number of unanswered questions for any investigation of land rights in the post-apartheid era, as well as raising the whole question of the meaning of security of title.

5

PERSONAL APARTHEID

Apartheid also operated at the micro level where separation affected the details of the daily life of the population. The policy was intended to eliminate virtually all personal contact between members of different population groups except within the master–servant or employer–employee relationship, both of which were essentially viewed in White–Black terms. Attention was focused upon the physical isolation of the White group from the others. For most personal apartheid matters there were only two salient groups: Whites and 'Non-Whites'. Because of the racial dualism which is evident in personal apartheid it is proposed periodically to refer to all who were not regarded as White as 'Non-Whites', however demeaning the term was considered to be. It illustrates the feeling of the intent of the multitude of laws and regulations issued to prevent those who were not classified as White from occupying and using declared White space. The term was widely used in everything from government notices to park benches.

Personal apartheid operated at various levels including the prevention of marriage between a member of the White group and a member of another group. It needs to be emphasized that the prohibition did not apply to intermarriage between the members of the Black, Coloured and Asian groups. In the occupation of space the laws operated from the level of separate park benches and entrances to buildings, to separate transport, schools, hospitals and ultimately, cemeteries.

A host of laws were introduced to enforce what was described as 'petty apartheid' as opposed to 'grand apartheid', involving political separation and urban residential segregation. Many of the laws had their predecessors in the pre-1948 era, but were made far more effective and comprehensive by the National Party government. Thus the Prohibition of Mixed Marriages Act (1949) and the Immorality Act Amendment Act (1950) were two of the earliest apartheid measures designed to preserve the imagined racial purity of the White group. Segregation was enforced within the domain. As a further measure, Black and Coloured servants living on White-owned properties were required to be provided with physically separate premises and entrances. Residential segregation was thus enforced within the confines of a single residential plot (Figure 5.1).

Separation at the scale of individual buildings was tightened. This ranged from the

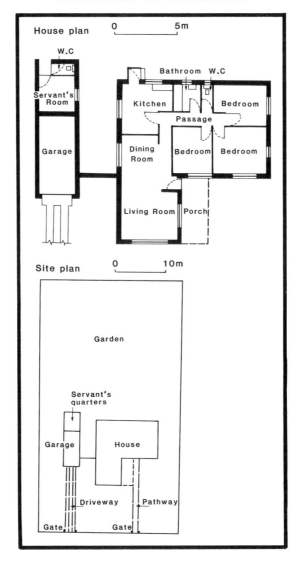

Figure 5.1 House with servant's quarters

Source Based on plans supplied by the Port Elizabeth municipality

provision of separate counters in post offices, to separate entrances to public buildings. Even separate stairways and elevators in multi-storey buildings were instituted where possible. Separate restrooms were also deemed essential. Such requirements had to be taken into account in the design of public and increasingly also of private corporate buildings. Thus the White and Non-White experiences of the same building were very different, as often separate offices dealt with the needs of the two groups. The plan of the Steytlerville post office illustrates the resultant duplication of entrances and public counters (Figure 5.2).

SEPARATE AMENITIES

The Reservation of Separate Amenities Act in 1953 together with its numerous amendments sought to create separate social environments for the White and other population groups. The Act was closely linked with the racial land zoning of the Group Areas Act, in order to determine who might use particular facilities. Accordingly parks, places of entertainment, dining and hotel accommodation were all regulated and reserved for a single group. Hence it became impossible for a White person to drink a cup of tea with a Non-White person in any cafe unless they obtained a permit to do so, which until the 1980s was unlikely to be issued. The grandiose term, 'International Hotel' was introduced in 1975 to allow Non-Whites with sufficient financial means to stay in selected, generally expensive, hotels. The cheaper establishments remained segregated. Regulations were such that within the hotels granted 'international' status, swimming pools and dance floors were initially denied to patrons who were not White, unless the swimmer or dancer could produce a foreign passport; and even then a Transkeian or Bophuthatswanan passport was not acceptable. In Durban it might be noted that the majority of the International Hotels were situated opposite a 'Whites-only' beach thereby leading to obvious disadvantages for Black patrons (Figure 5.3).

'Separate' under the Separate Amenities Act did not mean equal. Where, for example, only one waiting room existed on a railway platform, the stationmaster was empowered to set it aside for the use of White people only. Only later might a second, Non-White, waiting room be built.

Possibly one of the most controversial issues was the question of beach apartheid, as most of the coast line fell outside proclaimed group areas. Owing to the problems of defining the space involved, the matter was only brought within the ambit of the Separate Amenities Act by an amendment in 1960, which permitted local authorities to reserve both beaches and the adjacent sea for the exclusive use of a single group. The resultant zoning was remarkably complex and inequitable. Although in the majority of resorts the main beaches were reserved for Whites, peripheral beaches were reserved for other groups. Hence in Port Elizabeth the key beaches south of the harbour were reserved for Whites and only those deemed less desirable as adjacent to the industrial area, too rocky, or too remote were zoned for the other groups (Figure 5.4). For this purpose the full classification of the Group Areas Act was adopted with the absurd anomaly of separate beaches for Whites, Coloureds, Chinese, Indians, Cape Malays and Blacks. Local authorities which provided such zonings were then under an obligation to construct changing and ablution facilities for each of the groups. However, it was assumed by local town planners that only White people wished to enjoy, rather than could afford, beach recreation and hence few amenities were provided for the other groups until the 1980s.

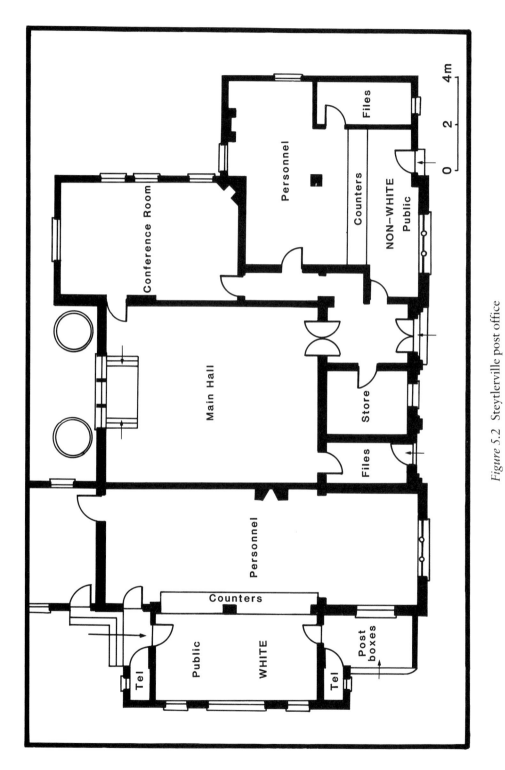

Figure 5.2 Steytlerville post office

Source After A.D. Herholdt (1986) *Die Argitektuur van Steytlerville met Riglyne vir Bewaring en Outwikkeling*, Port Elizabeth: Institute of South African Architects

Figure 5.3 Hotel classification in Durban, 1985

Source Based on information supplied by the Automobile Association

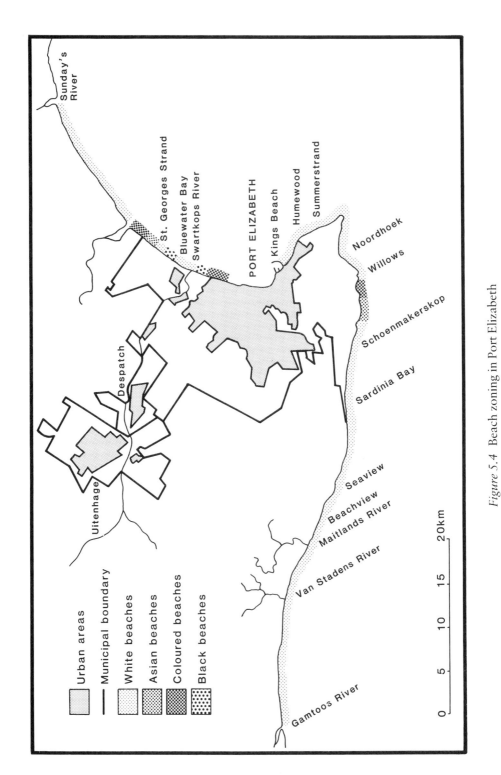

Figure 5.4 Beach zoning in Port Elizabeth

Source After A.D. Gibbon (1977) *Outdoor Recreation Survey of the Port Elizabeth Area*, unpublished MA thesis, University of Port Elizabeth

Legend:
- Urban areas
- Municipal boundary
- White beaches
- Asian beaches
- Coloured beaches
- Black beaches

Labels on map:
Sunday's River, St. Georges Strand, Bluewater Bay, Swartkops River, PORT ELIZABETH, Kings Beach, Humewood, Summerstrand, Noordhoek, Willows, Schoenmakerskop, Sardinia Bay, Seaview, Beachview, Maitlands River, Van Stadens River, Gamtoos River, Despatch, Uitenhage

Scale: 0 5 10 15 20km

Segregation was applied to sports clubs, sports grounds, swimming pools and other public open spaces. Sports facilities were governed by the Separate Amenities Act, even if privately owned. These included such major users of land as golf courses, which even if not within a proclaimed group area, were governed by the race of the owner. In this manner Whites could not compete against members of another population group in any sporting activity. Although participation remained segregated, some relaxation of the law was possible at certain sports stadiums and at race courses, where segregated enclosures for Whites and Non-White spectators were provided.

RELIGION

The government sought to introduce separation into church congregations, recognizing the importance of the religious community in social integration. In many respects the Group Areas Act effectively broke up integrated church congregations as disqualified people were removed from living in the vicinity of the church. Some continued to attend their original place of worship, even if this involved travelling considerable distances, as a means of registering their protest. The politically dominant church was fragmented into four separate entities. The Dutch Reformed Church (Nederduits Gereformeerde Kerk) remained for Whites, while the Dutch Reformed Mission Church (Nederduits Gereformeerde Sending Kerk) catered for Coloureds, the Reformed Church in Africa for Indians and the Dutch Reformed Church in Africa (Nederduits Gereformeerde Kerk in Afrika) for Blacks. The distribution of such churches followed the group areas pattern with few exceptions (Figure 5.5).

Other denominations did not follow the same approach and the church–state confrontation over the issue resulted in the *status quo* being maintained. Thus the Anglican Church of the Province of Southern Africa remained nominally integrated, although congregations tended to reflect the segregation of the residential areas. It does, however, also include the Order of Ethiopia, which is exclusively Black and operates outside the general parochial system. The host of separatist Black churches which came into being in the nineteenth and twentieth centuries represented a blending of African and European missionary traditions, which could not be accommodated within the Eurocentric churches. Such separation is therefore rooted in colonial attitudes rather than apartheid.

Spatial segregation continued even in death. The ultimate aspect of separation in social contact was the maintenance of segregated cemeteries. The majority of cemeteries in the nineteenth and early-twentieth centuries were segregated by religious denomination, often with sectors designated for individual churches. This allowed for some measure of legal population group integration where churches were open to everyone, but racial segregation where congregations were segregated. Areas within cemeteries not designated for the religious denominations were usually segregated with descriptions such as European 'free ground'. However, after 1950 new municipal cemeteries were zoned exclusively for the different population groups, where this was

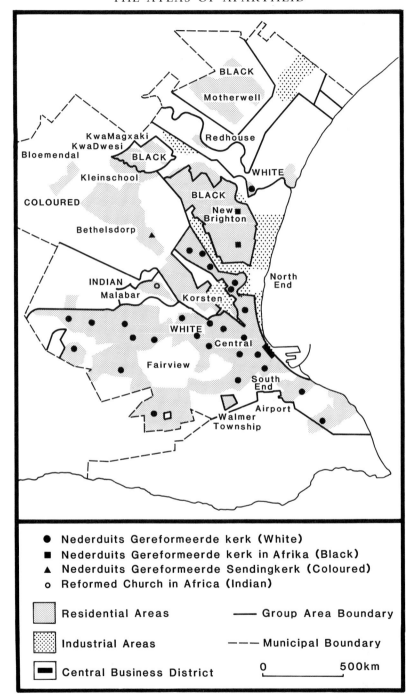

Figure 5.5 Dutch Reformed Church congregations in Port Elizabeth, 1990

Source Based on information supplied by the Dutch Reformed Church

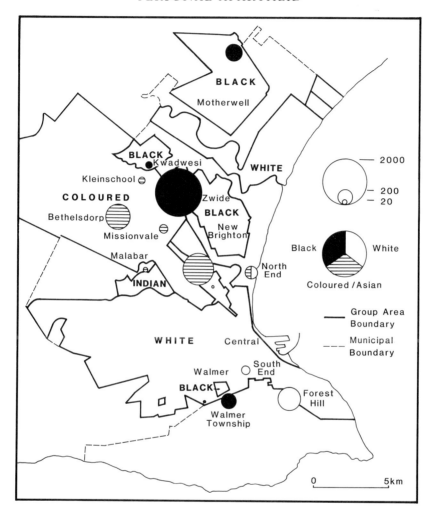

Figure 5.6 Burials in Port Elizabeth cemeteries, 1990/91

Source Based on information supplied by the Port Elizabeth, Ibhayi and Motherwell municipalities and the Cape Provincial Administration

not already done. In the main it was achieved in conformity with group areas zonings, although the older cemeteries continued to be used where families had already purchased burial plots (Figure 5.6). The pattern of separate burials is likely to persist, with planning for separate burial grounds continuing after the repeal of the Separate Amenities Act.

SEPARATE TRANSPORTATION

The enforcement of the Group Areas Act and the Natives (Urban Areas) Act ensured that movement between place of work and place of residence was usually separate, as a result of the different destinations to which the various groups travelled when going home. However, detailed configurations of bus routes and bus stops were carried over to the provision of separate sections of trains and buses for Whites and Non-Whites. The provision of separate buses resulted in separate routes, bus stops and termini. In a number of cities and on the inter-city bus services run by the South African Railways, separate sections of the bus were reserved for the two groups, necessitating the use of common routes, but separate shelters, ticket offices etc. In the railway system, trains were divided into White and Non-White sections, each separated for class, although Third Class was only provided for Non-Whites. This arrangement necessitated the provision of separate sections of railway platforms, each with an assemblage of ticket offices, waiting rooms and bridges over or tunnels under the tracks and an effective no man's land between the two.

Buses and trams presented a greater difficulty for the enforcement of segregation than trains, as everyone had to use a common entrance. Segregation had been imposed before 1948 in many services outside the Cape Province through the allocation of certain sections of a bus, usually the upper deck or the rear, to Non-White passengers and the remainder to Whites. In other centres separate buses and trams were run for Whites and Non-Whites. Some relaxation was locally permitted for Coloureds and Indians to use White facilities. In 1955 the government began to enforce regulations in those cities with integrated services, notably Cape Town and Port Elizabeth. At first (late-1950s) segregation on buses was imposed, with the reservation of different sections for the two groups, later (late-1960s) separate buses were enforced (Figure 5.7). Bus segregation was shortlived as in the late-1970s it was relaxed and finally abolished even in those cities where it had been firmly entrenched before 1948.

SCHOOL EDUCATION

The provision of education had been remarkably segregated before 1948. However, the new government was committed to the institution of Christian National Education, one of the tenets of which was the promotion of ethnic identity and solidarity. In 1949 the government appointed a Commission to enquire into the future of Black education. The result was the formulation of a new separate State education system for the Black population. It was also to be specific to Black needs and not, in Dr H.F. Verwoerd's terms, for the creation of 'expectations in life which circumstances in South Africa do not allow to be fulfilled'.[1] Under the new government Blacks were not to aspire to certain positions in society and so education for such positions was not deemed necessary. Control over Black education was taken by the Department of Native Affairs in 1954 under the Native Education Act, as the previous mission schools were regarded

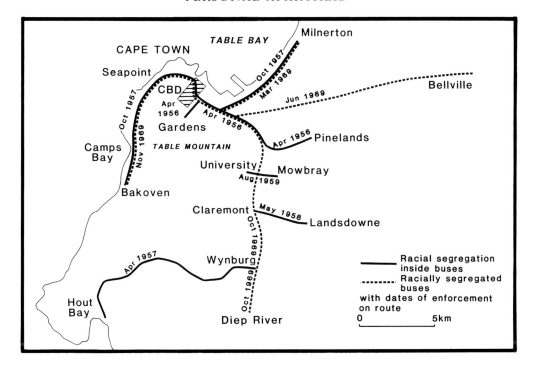

Figure 5.7 Racial segregation on bus routes in Cape Town

Source After G.H. Pirie (1990) *Racial Segregation on Public Transport in South Africa 1877–1989,*
unpublished Ph.D. thesis, University of the Witwatersrand

as fostering ideas, such as equality, which could not be encouraged.

One of the vital provisions of the new Black education system was the provision of mother tongue instruction, at least at primary level. The object was to fragment the Black population into separate nations, each with its own history and culture acquired through the education system. This requirement was the basis whereby different linguistic groups in the Black suburbs were to be housed separately, and provided a means whereby control over education could be linked to the Black homelands.

White schools were already separate, under the control of the four provincial authorities. Separate English and Afrikaans medium schools were provided throughout the country, together with such German schools as local communities could support. Dual or parallel medium schools were not encouraged in the interests of maintaining Afrikaner cultural identity. Provision was also made for separate Indian and Coloured education systems. Even exclusive Chinese schools were established in Port Elizabeth, in line with the Chinese group area. Only the private schools, including those run by church bodies, remained outside the direct control of the government. However, these schools were effectively run until the 1980s as segregated entities in order to obtain state recognition. The numbers attending such schools were small in comparison with

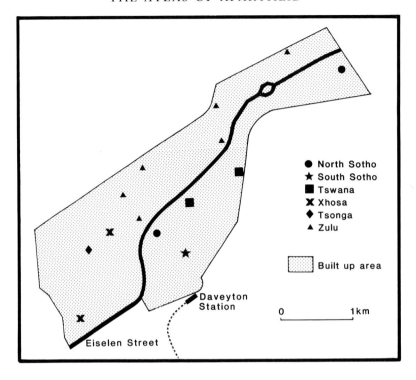

Figure 5.8 Primary school mother tongue, Daveyton

Source Based on information supplied by the Department of Education and Training

the total. Thus in the school sphere segregation was almost complete, exceeding that in residence (Figure 5.9).

In view of the significance of education in fostering the overall apartheid system, the control of Black education in the homelands was passed to the homeland governments. Accordingly by the late-1980s there were some seventeen departments of education in the country, in addition to a central national ministry.

UNIVERSITY EDUCATION

University education was similarly segregated. The Extension of University Education Act (1959) provided for the establishment of a series of new ethnically-based institutions for Blacks, together with separate universities for Coloureds and Indians, despite strong protests from the established universities. In seeking to create the new institutions the government indicated that 'they must be the bearers of their own national culture to stimulate that culture amongst their own national group' and that 'the future leaders should be educated and trained there, not to break down the colour bar but to retain it in the best interests of both Whites and non-Whites'.[2]

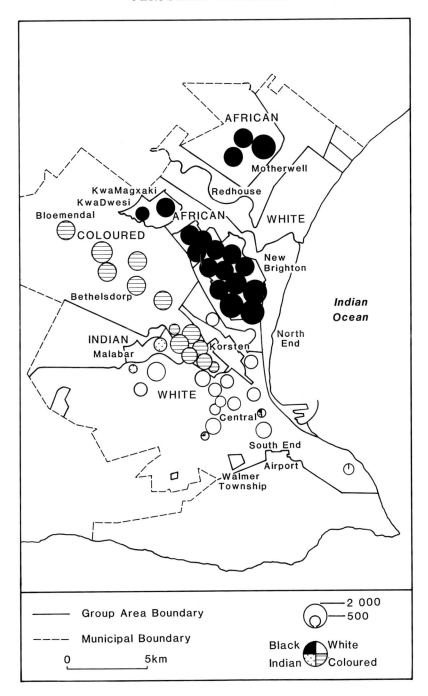

Figure 5.9 High school enrolments in Port Elizabeth, 1990

Source Based on information supplied by the various education authorities and private schools

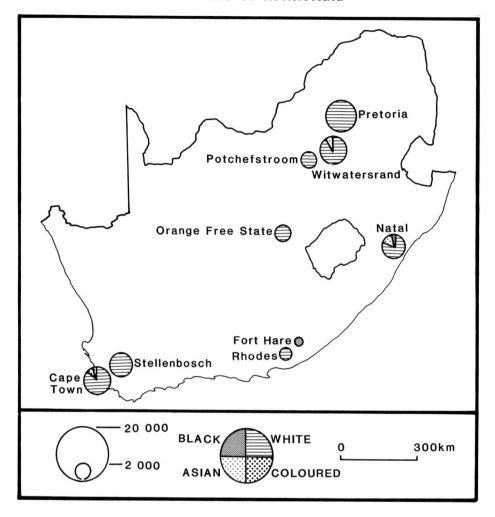

Figure 5.10 University enrolments, 1958

Source Based on statistics in South Africa (1961) *State of the Union 1960–61*, Johannesburg: Da Gama

In 1958 only 1,402 Asian, Black and Coloured students (17 per cent of the total) had been registered at university institutions classified as White (Figure 5.10). Thereafter only Whites were admitted to the existing universities, with the exception of Fort Hare, which was taken over by the government as an institution for the Xhosa population. The only other exceptions were the medical school of the University of Natal, which was open to Asian, Black and Coloured students only, and the correspondence institution, the University of South Africa. A few students, who wished to pursue a course not offered in their own ethnic university were permitted to register elsewhere. By 1974 when the system was in full operation only 2 per cent of the 78,000 students attending the residential universities in South Africa were registered at institutions for a designated population group other than their own. As in 1958, the largest such group was attending the University of Natal medical school (Figure 5.11). Academic staff were also race defined as only Whites could obtain posts at White universities, although the other universities were also staffed predominantly by Whites.

The ethnic exclusivity of the Black universities was abolished in 1979, when, for example, the University of Zululand was permitted to admit Black students other than those speaking Zulu or Swazi. Finally, in 1985 all universities were permitted to admit

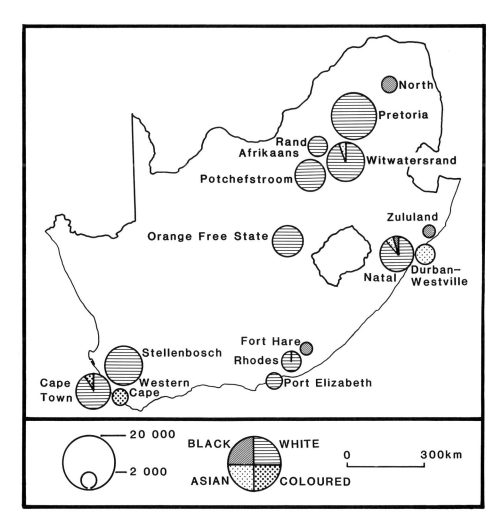

Figure 5.11 University enrolments, 1974

Source Based on statistics in South Africa (1975) *State of South Africa 1974*, Johannesburg: Da Gama

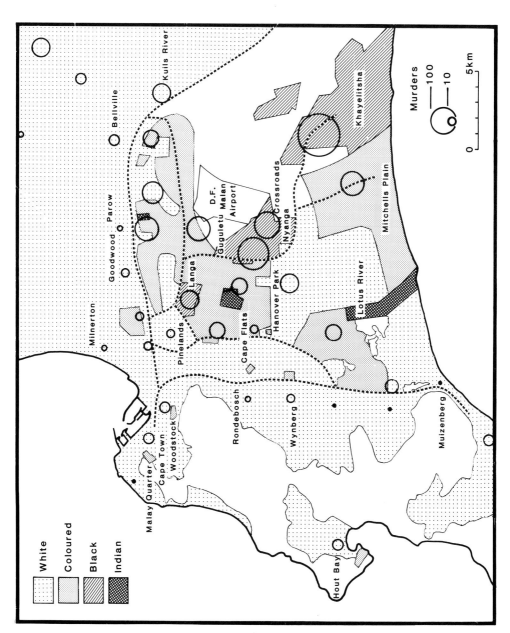

Figure 5.12 Murders in Cape Town, 1989

Source Based on statistics in *Hansard*, 21 March 1990, cols Q615–16

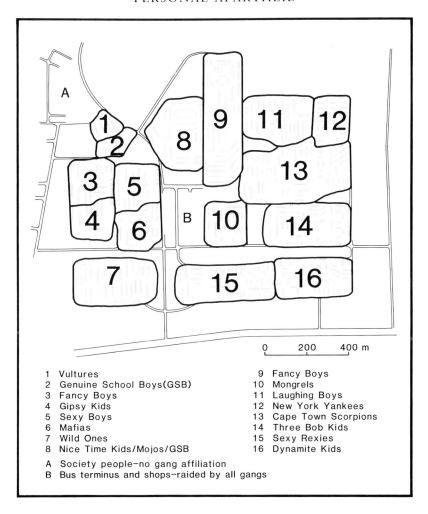

Figure 5.13 Gang territories in Hanover Park

Source After D. Pinnock (1984) *The Brotherhoods: Street Gangs and State Control in Cape Town*, Cape Town: David Philip

any student regardless of population group, marking one of the earliest breakdowns in the apartheid system.

The homeland policy was also responsible for the creation of yet more universities. The newly independent states, Transkei, Bophuthatswana, and Venda, all established new universities as one of the status symbols of separate statehood. Qwaqwa also set up a branch of the University of the North to which South Sotho-speaking students had previously been directed. Thus by the early-1980s there were separate residential universities for virtually every Black linguistic group.

More institutions were established to supplement the existing facilities. In 1976 the

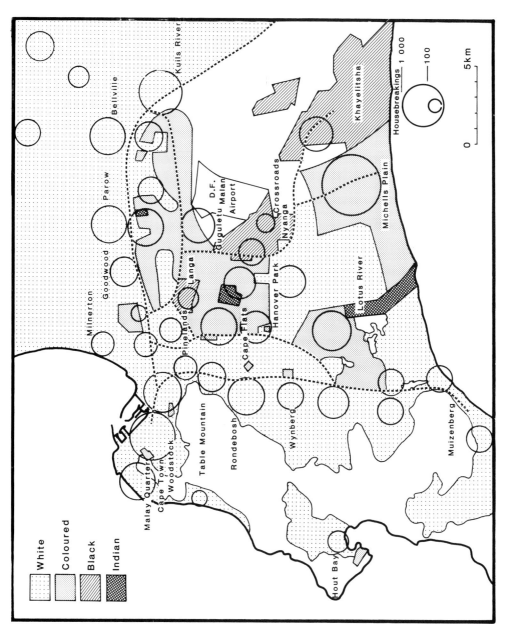

Figure 5.14 Housebreakings in Cape Town, 1989

Source Based on statistics in *Hansard*, 21 March 1990, cols Q615–16

White
Coloured
Black
Indian

Housebreakings
1 000
100

0   5km

Milnerton
Goodwood
Parow
Bellville
Kuils River
Khayelitsha
D.F. Malan
Airport
Guguletu
Crossroads
Nyanga
Langa
Pinelands
Cape Flats
Hanover Park
Michells Plain
Lotus River
Malay Quarter
Cape Town
Woodstock
Table Mountain
Rondebosch
Wynberg
Muizenberg
Hout Bay

Medical University of South Africa was created outside Pretoria for the training of additional Black doctors. In 1983 Vista University was established to provide urban campuses for the Black population, where no branch of a Black university existed in suburbs under homeland control. The aim of the institution was to provide tertiary education for Black students in the larger cities, outside the western Cape. As such it marked a major departure from the previous approach of sending students for further education back to the homelands in order to foster national identities.

PERSONAL SECURITY

It is remarkably difficult to evaluate spatially the various indicators of personal well-being as most were not made available for small area analysis. However, some do lend themselves to geographical analysis when a statistical source is offered on a spatial basis. The differing levels of security illustrate aspects of living in a divided city through the availability of police station records, offered in response to questions raised by members in parliament.

The Cape Town metropolitan region has been taken as an example. In 1989 the region recorded 1,483 murders. This indicated the exceptionally high rate of approximately 70 murders per 100,000 population. The distribution of reported murders exhibited a marked concentration in the Coloured and Black suburbs (Figure 5.12). This has several explanations related to the extent of socio-economic deprivation within the two communities, and to the concentration of policing in the White sections of the region. The particularly high number of murders in the Khayelitsha squatter area is indicative of this relationship. In contrast the White suburbs on the western and northern sides of the metropolitan area recorded markedly fewer murders.

In some of the displaced communities on the Cape Flats the levels of gang warfare achieved frightening proportions as gangs struggled for turf in some of the more notorious suburbs (Figure 5.13). The remarkably high murder rates among young Coloured males has attracted considerable attention, with such headlines as 'even the Grim Reaper is not colour-blind'.[3] In the age group 15 to 24, Coloured males are almost ten times as likely to be murdered than White males and twice as likely as Black males.

On the other hand when reported housebreakings are examined the pattern is changed (Figure 5.14). Central Cape Town, which listed few murders, recorded the highest levels of housebreakings. Profitable suburban burglary in the White areas gave rise to intense activity by the security industry, and a change in the appearance of many streets as high walls with security gates replaced low walls and hedges around houses. Furthermore, in the Coloured areas the comparatively affluent new suburb of Mitchell's Plain recorded the highest numbers of housebreakings. The Black suburbs reported the lowest levels, suggesting either low levels of reportage or little economic incentive for the activity. Thus within the apartheid city the various communities had, as with so many aspects, markedly different security priorities.

6

RESISTANCE TO APARTHEID

THE BUILD-UP OF RESISTANCE

In 1948 the Government embarked upon a series of wide-ranging new policies, which encountered widespread opposition, notably from the Black population which was most severely affected. The initial stage of opposition was the Programme of Action, when the policy of deputations and petitions was rejected in favour of boycotts, strikes and civil disobedience. However, opposition was not united and the Durban riots in January 1949, when 142 people were killed, illustrated the tensions between the Indian and Black communities. In 1952 the joint Defiance Campaign was launched to obstruct the imposition of the multitude of repressive laws which were emanating from parliament. To counter opposition, the Suppression of Communism Act of 1950 was passed, with a remarkably flexible definition of what constituted the promotion of that ideology. Confrontation over the banning of the South African Communist Party led to more deaths, and the virtual absorption of the party's programme and personnel by the African National Congress. This was symbolized by the adoption of the Freedom Charter in 1956, to the unease of the Africanists, who seceded to form the Pan Africanist Congress in 1959. Defiance campaigns and legal contests remained the main form of opposition throughout most of the 1950s, while the government plugged the loopholes in the legislative and administrative structures of apartheid, and adopted further security legislation to enforce them.

The introduction of new rural Black local authorities legislation, agricultural betterment schemes and forced removals in the later 1950s led to physical confrontation in a number of regions, notably in the Transvaal in Sekhukuneland and the Hurutshe and Mamatlhola's Locations. In Pondoland in 1958–60 a state of insurgency developed, culminating in military action against organized rural rebels in the taking of Ngquza Hill (Figure 6.1).

The development of official policies, notably separate education, passes for Black women, and the large-scale removals resulted in general unease and resentment. When added to the continuous problems of poverty, squatter removals, liquor raids, and bureaucratic intransigence, the potential for violence within the Black communities was

161

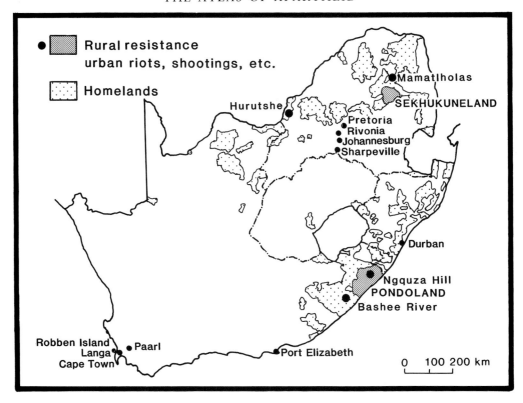

Figure 6.1 Early resistance to apartheid

Source: Compiled by the author

high. In 1959 riots took place in Durban's Black townships, and in the following year nine policemen were killed after a liquor raid in Cato Manor in Durban. Both the African National Congress and the Pan Africanist Congress organized demonstrations in connection with a variety of grievances, more particularly the pass laws. On 21 March 1960 at a Pan Africanist Congress pass law demonstration at Sharpeville in the southern Transvaal, sixty-nine Black people were killed when fired on by police, initiating national and international protest. More protestors were killed at Langa in Cape Town on the same day.

The immediate government responses were the imposition of a state of emergency, and the banning of both the African National Congress and the Pan Africanist Congress. Both organizations thereupon formed military wings, Umkhonto we Sizwe (Spear of the Nation) and Poqo (Standing Alone) respectively, to attack installations and officials and seize power by force. Riots in Paarl and the murders at Bashee River prompted a decisive response. The government was able within a couple of years to effectively crush such opposition, imprisoning the principle leaders of both organizations, and exiling the remainder. The arrest of the African National Congress leadership at Rivonia in 1963 proved to be decisive. The high security prison on Robben Island off Cape Town accordingly held the leaders of all the major anti-government factions until their release in 1989–90.

ARMED RESISTANCE

The exiled organizations established links with a number of countries for the training of their personnel. Consequently camps were established in Tanzania and later in Angola, Mozambique and Zambia for the instruction of the military wings of both the African National Congress and the Pan Africanist Congress (Figure 6.2). Specialized training was undertaken in a number of sympathetic countries, notably in the Soviet Union and its allies. The headquarters of the African National Congress were established in Dar es Salaam, but moved to Lusaka after the Morogoro conference in 1969, in order to be closer to South Africa.

After 1976 the African National Congress was faced with the task of sustaining a substantial number of refugees from South Africa. The number of exiles increased from possibly 1,000 in 1975 to 9,000 five years later. In order to accommodate this increase a new centre was established at Mazimbu, near Dar es Salaam, in 1979. Education and training was provided at the Solomon Mahlangu Freedom College. The period after 1975 was one of reorganization as bases were established closer to South Africa in an attempt to intensify armed resistance. Angola became the main African National Congress military centre with bases in areas under the control of the Angolan central government. The attempt to establish bases in Mozambique was temporary as the 1984 Inkomati Accord led to their removal. Similarly, activity in Zimbabwe was ended with the election of the Zimbabwe African National Union–Patriotic Front government in 1980, which backed the pan Africanist Congress. Thus the African National Congress activity was largely confined to the three sanctuary states of Angola, Zambia and Tanzania. Its resources were spread widely in order to prevent undue influence being exerted by any one state. Understandably little comprehensive information is available on the activities of the bases, force numbers, etc. while the possibility of renewed conflict exists.

1976 SOWETO RIOTS

By 1976 the long period of apparent security force supremacy had begun to wane. The independence of Angola and Mozambique under Marxist governments and the increasing South African military involvement in both, brought the prospects of Black empowerment closer. The authorities on the other hand pursued their policies with increasing insensitivity thereby breeding intense resentment. The point of conflict came over the imposition of new regulations attached to the Bantu Education programme, which sought to introduce the compulsory use of Afrikaans as the medium of

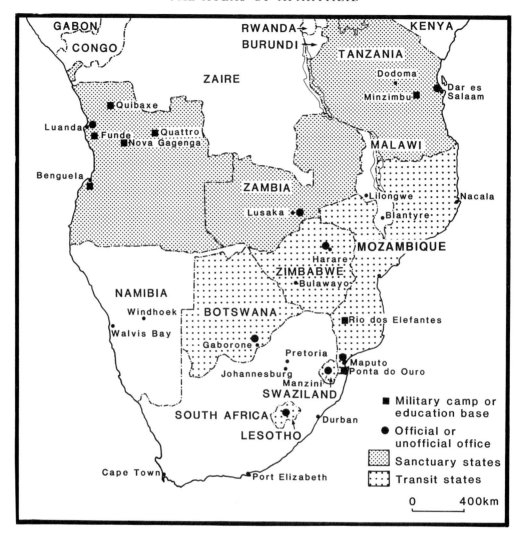

Figure 6.2 African National Congress bases

Source After S.M. Davis (1987) *Apartheid's Rebels*, New Haven: Yale University Press

instruction for mathematics, social studies, history and geography. The burden of this requirement, added to the perceived inferiority of the programme itself, resulted in protests and demonstrations, organized by the South African Students' Movement.

June 16th 1976 was chosen as a day for country-wide demonstrations against the imposition of Afrikaans in Black schools. The involvement of many strands of discontent allowed the whole situation to lead to riots. A march by school children in Soweto ended in confrontation with the police as the necessary permission from the

164

West Rand Bantu Affairs Administration Board had not been given. The police opened fire killing a number of young demonstrators. Serious riots broke out the following day with associated murder, arson and looting. The particular targets of the arsonists were government properties, notably those administered by the Bantu Affairs Administration Boards, including libraries, clinics and beerhalls. Schools, shops and houses were also attacked.

The riots spread from Soweto to other towns on the Witwatersrand as well as to Pretoria and its vicinity (Figure 6.3). Extensive damage was done to the Universities of the North and Zululand. It was only in August that major rioting began in the western Cape, when it had largely died down on the Witwatersrand. The Black areas in Cape Town were affected in mid-August and the Coloured areas in early September. Riots in Port Elizabeth in mid-August were similarly directed against Administration Board property. Sporadic outbreaks occurred elsewhere in the country but were not as sustained as those on the Witwatersrand and Cape Town. Almost half the 575 deaths occurred in Soweto alone. Natal and the central areas of the country experienced little serious threat to the state. The government was able to restore its authority, and

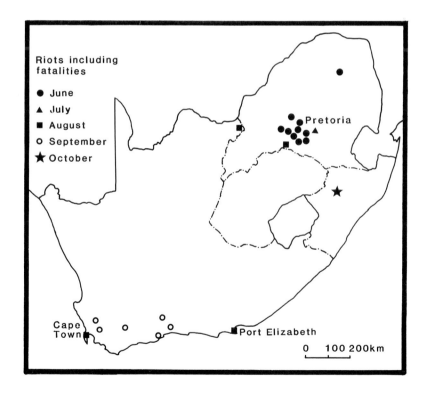

Figure 6.3 The diffusion of violence, 1976

Source: From information in South Africa (1980) *Report of the Inquiry into the Riots at Soweto and Elsewhere*, Pretoria: Government Printer

embarked on a limited reform policy. For several thousand young Blacks, however, the riots provided the opportunity to flee the country and join the various exiled organizations seeking the overthrow of the South African government. The role of pupils and students had been of major significance in the origin and continuation of the riots and breakdown of authority, whether state or parental.

1984 AND THE 1985–90 STATES OF EMERGENCY

The introduction of the new constitution in 1984 with the structural exclusion of the Black population from the decision-making process of the country, led to a general state of unrest ending the comparative quiescence of the early-1980s. In September 1984 riots broke out in a number of Black townships in the southern Transvaal, notably in the Vaal Triangle. The death rate due to political violence increased sharply. In addition to the constitution, rent and transport increases, together with a school boycott, created an explosive environment. Discontent with the new structures of Black local government held responsible for many of the townships' ills, also provided a visible target for the riots. Meetings were banned and the army was widely deployed in support of the police in controlling demonstrations, searching houses for arms and maintaining order. Funerals became the main outlet for political gatherings.

Boycotts and demonstrations spread to other parts of the country in early 1985, notably to the eastern Cape. Furthermore, much of the unrest was experienced in the smaller towns, such as Cradock, Colesberg and Graaff-Reinet. On 21 March 1985, the twenty-fifth anniversary of the Sharpeville shootings, police opened fire and killed twenty mourners in a funeral procession in Langa township, Uitenhage. The incident was highly inflammatory both internally and internationally. Violence subsequently became particularly deadly in a number of centres on the East Rand.

A State of Emergency was declared in July 1985 in several eastern Cape and East Rand districts (Figure 6.4). Wide powers of arrest and detention were given to the police. Major drives against anyone suspected of violence or incitement ensued, although the majority so detained were released quite rapidly, owing to the exceptionally large numbers involved. Press censorship was imposed, thereby hiding the activities of death squads with possible police and army connections, only now being revealed by the investigative journalism of the alternative press. The emergency was lifted in a number of districts in October. However, violence spread to the western Cape and a state of emergency was imposed there in the same month. The emergency was lifted in more districts in December, and in its entirety in March 1986.

The 1984–6 period was preceded by raised expectations on the part of the Black population as the reform process appeared to offer prospects of change. In 1983 the United Democratic Front had been launched to oppose the racialization of the constitution. However, Blacks were excluded from the 1984 constitution which provided only for White, Coloured and Indian houses of parliament. Furthermore, the economy was failing to grow at a rate to provide jobs for school leavers or provide any

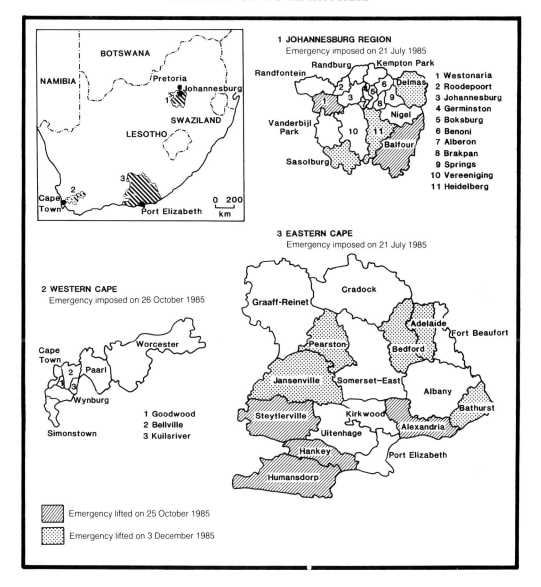

Figure 6.4 Extent of the state of emergency, 1985

Source: After A. Lemon (1987) *Apartheid in Transition*, Aldershot: Gower

material improvement in living conditions. The Black political parties and trade unions, newly legalized in 1979, thus had ample ground upon which to work to gain support.

Levels of violence continued to be high and a further state of emergency was imposed on the entire country in June 1986 (Figure 6.5). A dramatic lessening of violence and politically motivated murder ensued, as massive detentions were carried out and the

167

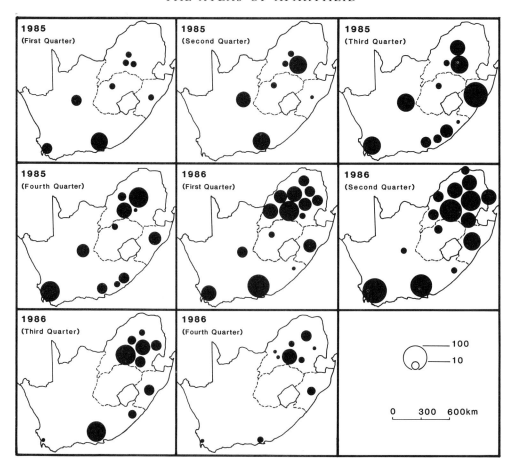

Figure 6.5 Deaths due to political violence, 1985–6

Source: Based on information supplied by the South African Institute of Race Relations

organizers of the various factions were scattered. Relative peace returned to the country, if under a considerable degree of repression. The year 1987 was one of the more peaceful in recent South African history, apart from the initiation of the massive civil strife in Natal and KwaZulu, which began in the second quarter (Figure 6.6). By 1989 the scale of the fighting and political murder in Natal exceeded that under the 1986 state of emergency for the country as a whole. The struggle between Inkatha and African National Congress-affiliated groups was such that the resources of the KwaZulu government and indeed other, then secret, sources had to be invoked to prevent the collapse of the Inkatha movement. Little headway was made by the police or the security forces in containing and preventing the violence, leading to suggestions of continuing police collusion with Inkatha.

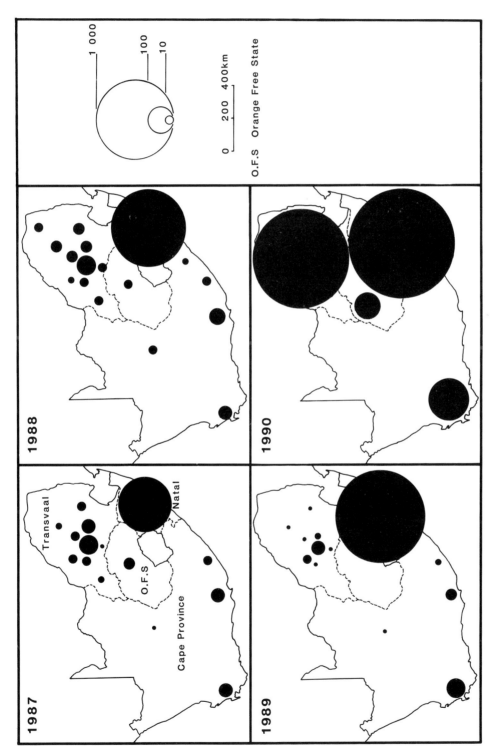

Figure 6.6 Deaths due to political violence, 1987–90

Source: Based on information supplied by the South African Institute of Race Relations

The wave of political turmoil associated with the removal of the ban on the African National Congress, the Pan Africanist Congress and the other movements, and the lifting of the state of emergency in 1990 resulted in an outburst of violence, which exceeded anything previously experienced under National Party rule, culminating in the death of nearly 700 people in the month of August 1990. Riots and killings spread rapidly to other parts of the country, although Natal remained the single greatest source of political murders. The conflict in the southern Transvaal involved the repetition of the Inkatha–African National Congress struggle as single men's hostel dwellers, mostly Zulu, battled with Congress supporters in the townships particularly on the East Rand. Over 800 people were killed in the course of the year. Conflicts took place between township dwellers and squatters, between different factions in the movements, the Pan Africanist Congress, and in Port Elizabeth in August 1990, an outburst involving the gutting of commercial premises directed against the Labour Party. These resulted in efforts on the part of all the political leaders to quell the violence, but to little avail as violence was only partly related to political party conflicts while other grievances were aired and old personal and ethnic scores were settled in the name of national political slogans. Political killings continued to run at the exceptionally high rate of approximately twelve a day in 1991 and 1992.

DEATHS IN DETENTION

As part of the State of Emergency, declared in 1960 following the Sharpeville killings, the government introduced detention without trial. In 1963 under the General Laws Amendment Act, later expanded as the Internal Security Act, this power became a permanent part of the State apparatus in fighting opposition. The period of detention was lengthened, and then made indefinite if sanctioned by a judge in 1966. Even such cursory authorization became unnecessary after 1976. The Transkei, Bophuthatswana, Venda and Ciskei governments introduced their own, often more draconian, legislation on attaining independence.

In the course of the period from 1960 to 1990 some 78,000 people were detained without trial. One of the notable features of this legislation was the virtually unlimited power exerted by the Security Police over the detainees, and the consequent number of deaths in detention (Figure 6.7). In the 1960s two or three detainees died each year, until the widely publicized death of Ahmed Timol in 1971. No more deaths were recorded until the Soweto riots of 1976. There then followed a horrific 26 deaths in two years. The international outcry over the much publicized death of Black Consciousness leader Steve Biko in 1977 again brought a halt to the deaths. However, in the 1980s numbers of deaths once more began to rise until the end of the final State of Emergency. Thus some seventy-three people died in detention in the period, approximately 1 person per 1,000 held. The figures do not include those who died in police custody but not specifically related to the enforcement of the security laws. The ruthlessness of the Security Police was encouraged by the ethos of the government from senior ministers

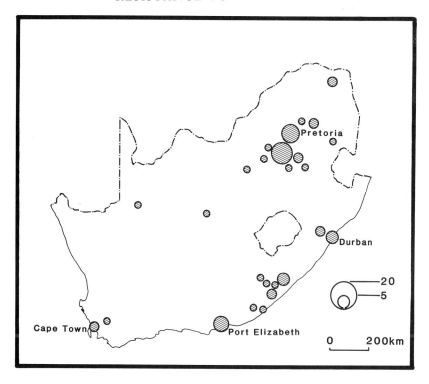

Figure 6.7 Deaths in detention, 1963–90

Source: Based on information supplied by the Human Rights Commission

downwards. The Minister of Police, J. Kruger's comment on the death of Steve Biko: 'I'm not pleased, nor am I sorry; Biko's death leaves me cold'; is eloquent testimony to the official lack of concern at the methods employed.[1]

The Security Police operated widely not only from their own regional headquarters, but also from local police stations in the smaller towns. Consequently although the major cities, such as Johannesburg (fourteen deaths) and Pretoria (ten deaths), dominate the list, the smaller centres and the homelands, particularly the Transkei, feature prominently. Individual buildings, notably John Vorster Square police station in Johannesburg (seven deaths), Pretoria Prison (five deaths) and Johannesburg Fort and the Sanlam Buildings in Port Elizabeth (both four deaths) feature prominently in the Security Police's drive against opposition. The southern Transvaal and the eastern Cape emerge as the most dangerous regions for political opponents of the government.

7

INTERNATIONAL RESPONSE TO APARTHEID

South Africa has gained dubious prominence in international politics during the post-Second World War era, through the almost universal condemnation of the policy of apartheid. However, several other related issues have further complicated the international relations of the state. Prominent among them have been the question of the South African administration of South West Africa (Namibia), which bedevilled relations between the United Nations and South Africa until 1990, and the expansionary intent of South African governments until the 1960s with regard to the then British High Commission territories (now Botswana, Lesotho and Swaziland). Further issues, notably economic and political sanctions imposed by other states against South Africa and the retaliatory destabilization of its neighbours by South Africa, led to a significant, and possibly decisive, impact of external affairs on the pursuit of apartheid.

In the regional context South Africa has operated as a local superpower exerting direct influence upon a wide range of countries in the southern cone of Africa through its economic strength (Figure 7.1). The Southern African Customs Union and the Rand Currency Area, the former dating from the time of the formation of the Union in 1910, have been important organizations for exerting economic influence. Furthermore, the configuration of the transportation system, notably the railway network built in colonial times, focuses intercontinental trade routes upon the harbours of South Africa. This position was exploited by the South African government in an attempt to obtain international recognition of its apartheid policies.

DIPLOMATIC ISOLATION

South Africa has been subjected to intensive diplomatic pressures. This is evident not only in the volume of hostile resolutions adopted by the United Nations, the Organisation of African Unity and other international bodies, but the country's exclusion from them. In addition there was a remarkable low level of foreign diplomatic representation within the country, and a limited range of South African official accreditation in other countries. The South African government thus maintained a diplomatic network far smaller than its size and power would have suggested.

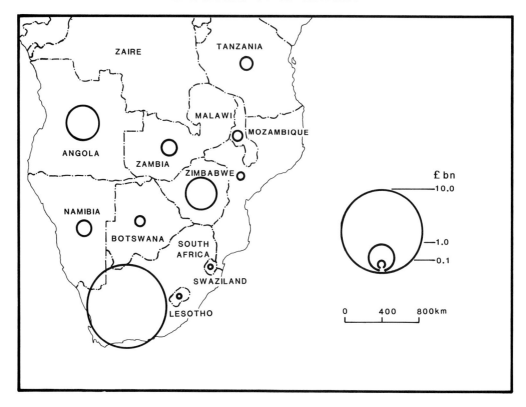

Figure 7.1 Southern Africa: Government expenditure, 1985–6

Source Based on figures from the *Statesman's Yearbook*, 1985–6

As a result of the condemnation of apartheid policies, the country was relegated to the status of an international pariah (Figure 7.2). This was a gradual process rather than a sudden occurrence. Thus South Africa maintained some twenty-one diplomatic missions abroad in 1955 and twenty-five in 1989, although the number of independent states worldwide had increased markedly in the period. The diplomatic missions maintained in Transkei, Ciskei, Bophuthatswana and Venda are excluded from this discussion. These four states remained totally isolated internationally except for mutual and South African recognition for the entire period of their existence.

In the late-1960s and early-1970s the South African government attempted to break out of this isolation by extending its network, mainly with military and fascist orientated regimes in Latin America. However, the gains were ephemeral as increasing pressures were exerted to isolate the country. The distribution of government missions was heavily concentrated in western Europe and represented long-term links. Most areas of the world were effectively closed to South African diplomats. Of significance was the lack of official relations with other African states, with the exception of

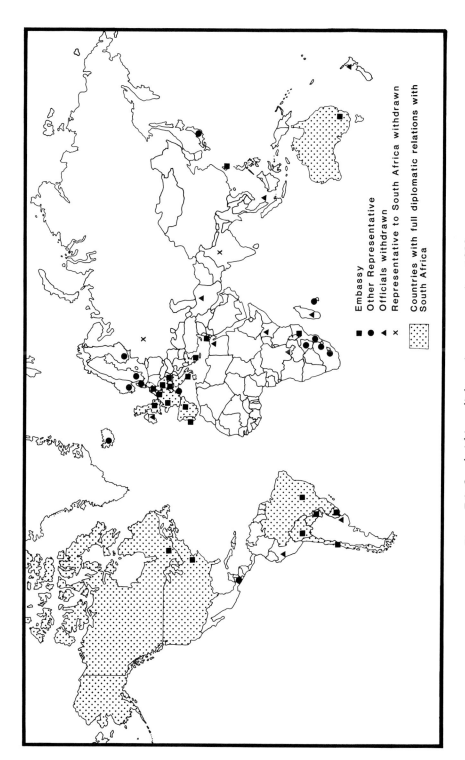

Figure 7.2 South African diplomatic representation, 1989

Source Based on South Africa (1990) *Official Yearbook 1989*, Pretoria: Government Printer

Malawi. In Asia diplomatic relations were only maintained with the two other international outcasts, Israel and the Republic of China (Taiwan). Elsewhere only the United States, Canada and a small group of Latin American states retained formal links.

In a number of cases lines of communication were maintained at a lower level, such as through a consul or trade representative, most notably with Japan. In certain instances these were the downgraded form of previous full diplomatic representation. In others they formed a new link, which owing to the general diplomatic climate could not be at embassy or legation level. Thus the consulate in Maputo ceased to function after the independence of Mozambique in 1975, but was reopened in 1984 following the Nkomati Accord, although full diplomatic relations were not established.

The low level of foreign contacts was partially self-imposed, as symbolized by the country's withdrawal from the Commonwealth in 1961. Relations with countries such as the Soviet Union were terminated at South Africa's insistence as part of its racial and anti-communist policies. While India terminated its representation as a result of the South African government's failure to show any flexibility over the disadvantaged position of the Indian community. However, the main cause of the loss of representation resulted from sanctions, mostly imposed at independence in the case of African states. The diplomatic efforts to solve the South West African independence issue, initiated with the protocol signed in Brazzaville in 1989, began the process of South Africa's re-entry into the mainstream of diplomatic representation.

AIRWAYS SANCTIONS

One of the most spectacular spatial examples of the imposition of sanctions was that applied to air communications with South Africa. In 1962 the United Nations called upon member states to refuse landing and overflying rights to South African aircraft. The reciprocal refusal to allow other airlines to fly to South Africa was taken subsequently by a number of countries as a logical extension of sanctions.

The main destination of travellers from South Africa was western Europe. Thus in 1963 a number of African states were able to take advantage of their strategic position and refused to allow South African Airways use of their airspace (Figure 7.3). This necessitated the re-routing of flights to Europe around the west coast of Africa, rather than use the previous more direct route via Nairobi. South African Airways was forced to absorb the increased costs of the longer route (approximately 1,400 kilometres) and, by using the Portuguese airports at Luanda and Ilha do Sal in the Cape Verde Islands, overcome the problems of restricted aircraft range. After independence Angola joined the boycott, but the Cape Verdian government was too dependent economically on the landing fees to do so. The introduction of longer ranged aircraft enabled South African Airways to reduce its dependence upon intermediate airports except in emergencies. Thus the later acquisition of landing rights at Abidjan in the Ivory Coast was the result of political, not operational policy. The extreme consequence of the ban on overflying

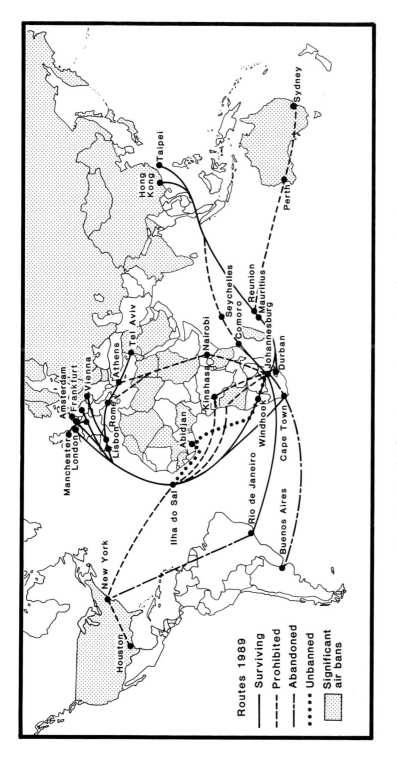

Figure 7.3 South African Airways intercontinental routes, 1989

Source After G.H. Pirie (1990b) 'Aviation, apartheid and sactions: air transport to and from South Africa, 1945–1989', *GeoJournal* 22: 231–40

was the routing of South African Airways flights to Tel-Aviv around the west coast of Africa.

Other intercontinental routings were more problematical, reflecting the country's external relations. The direct air link with the United States proved to be politically the most contentious. Inaugurated in 1969, the flights were the subject of controversy from the beginning. However, it was only in 1986 with the major thrust of the Comprehensive Anti-Apartheid Act that bans were placed on South African Airways and American airlines from operating between the two countries.

Links across the Indian Ocean were equally problematical and politicized. The Australian airline, Quantas, began operating between Sydney and Johannesburg as part of the Commonwealth linkup programme in 1952, and South African Airways began a reciprocal service five years later. However, the political pressures of the post-1984 era were such that in 1986 the Australian government also prevented further direct flights by both airlines between the two countries. Quantas then flew to Harare, necessitating inconvenient stopovers for South African passengers, while South African Airways flights to Hong Kong provided an alternative routing.

South African Airways established routes to Hong Kong and Taipei, via Mauritius, but direct links with Japan and India eluded the corporation. In South America, links with Brazil and Argentina were established, but the latter were withdrawn as a result of unprofitability, although the collapse of the military dictatorship in Argentina undoubtedly contributed to the decision, as symbolized by the downgrading of diplomatic relations.

Within the southern African region, links with Zimbabwe were retained at independence, as were those with Malawi, Zambia, Mozambique, Botswana, Lesotho and Swaziland. In the Indian Ocean, Mauritius retained links, while the Seychelles banned flights in 1980, only to recommence them in 1990 as part of the ending of air sanctions. Homeland airlines, including Transkei Airways and Ciskei International Airlines failed to establish any links outside South Africa.

Sanctions were thus effective in reducing the overseas range of South African Airways operations. However, they continued, and in many cases increased numbers of flights offered by a wide range of European airlines effectively enabled links to be maintained with the rest of the world.

THE HIGH COMMISSION TERRITORIES

Until South Africa withdrew from the Commonwealth in 1961, one of the constant aims of successive governments had been the incorporation of the three British protectorates of Basutoland, Bechuanaland and Swaziland. Provision had been made in the British South Africa Act of 1909, which established the Union of South Africa, for the eventual inclusion of the three territories, as well as Southern Rhodesia (Zimbabwe) (Figure 7.4). The electorate of the latter territory in a referendum in 1923 decided not to join the Union. Various schemes of incorporation of the three remaining territories

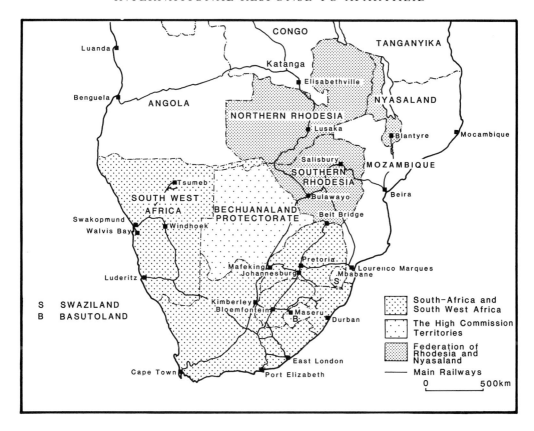

Figure 7.4 Southern Africa, 1961

Source Compiled from *University Atlas* (1962) George Philip: London

were negotiated in the period up to 1948. However, they all failed to materialize as successive British governments, of whatever persuasion, remained suspicious of South African intentions towards the indigenous population and the preservation of its land rights.

The National Party government after 1948 remained equally committed to incorporation. Indeed official reports, including that of the Tomlinson Commission, drew maps and published statistics effectively including the three High Commission territories into the Black areas of South Africa (see Figure 3.2). Accordingly it was possible to suggest that the indigenous population had retained half the land area of 'greater' South Africa. The three protectorates thus were initially viewed as part of the grand apartheid design.

Economically they operated as integral parts of the country through the currency and customs union and, until 1963, the absence of most forms of control over the borders between the states. Although South Africa failed to gain sovereignty over the countries, it was able to exert influence over them, as witnessed by the successful pressure applied over the marriage of Sir Seretse Khama in 1949, leading to his removal from the chieftainship. However, once South Africa had left the Commonwealth the possibility of political incorporation was not considered again and the three states became independent in the mid-1960s.

In 1954 the British government launched the Federation of Rhodesia and Nyasaland on the political platform of 'partnership' between Europeans and Africans. In this respect it was designed as a counter to South African influence in the region and a demonstration that there was a better alternative to apartheid. The new federation lessened its economic dependence upon South Africa through such schemes as the direct rail link to Lourenço Marques and the promotion of manufacturing industry.

SOUTH WEST AFRICA

The legal relationship between South Africa and South West Africa was one of the most significant international problems to confront successive South African governments between 1915 and 1990. In 1915 South African forces occupied the German colony. In 1919 the League of Nations awarded South Africa a mandate to administer the territory on its behalf. The mandate was held under class C, which placed few restrictions upon the freedom of the mandatory power to rule as it thought fit, as no concepts of self-government were written into this category of trusteeship. The South African government was allowed to administer the territory as an integral part of the Union, limited only by an obligation 'to promote to the utmost the material and moral well-being and social progress' of its indigenous inhabitants who were largely confined to a series of reserves (Figure 7.5).[1] However, with the benefit of proximity the government was able to encourage South African, mainly Afrikaner, land colonization and the extension of White control on the same lines as in South Africa.

After the Second World War the League of Nations mandated territories were transferred to the trusteeship of the United Nations. The South African government refused to make the transfer because greater obligations, notably those of indigenous political advancement, were imposed. Instead the government sought to incorporate the territory into South Africa, but this course of action was rejected by the United Nations in 1946. The status of the territory thus became a matter of bitter dispute, with the United Nations attempting to exert influence over the activities of the mandatory power, while South Africa fended off what it considered to be interference in its internal affairs. After 1948 the government refused to render any more reports in terms of the mandate and the following year took a step nearer annexation by providing for the representation of the territory's White population in the South African parliament. In 1966, after lengthy proceedings at the International Court of Justice at the Hague, the United Nations revoked South Africa's mandate. During the following year a Council for South West Africa was established, under a Commissioner, charged with the task of preparing the territory for independence. In 1968 the international body symbolically

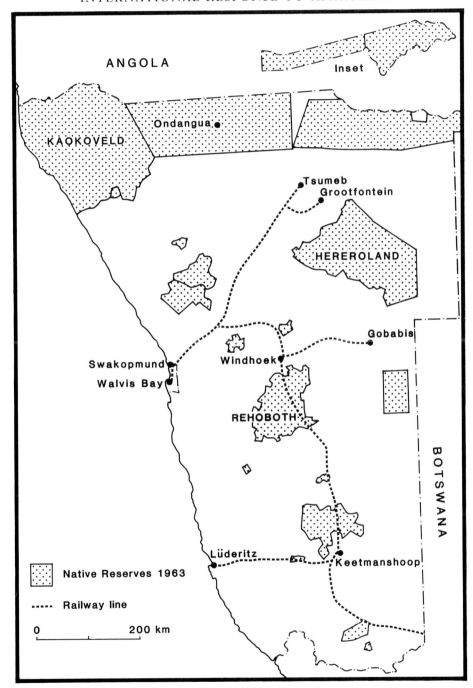

Figure 7.5 South West Africa Native Reserves

Source After South Africa (1964) *Report of the Commission of Enquiry into South West African Affairs 1962–63*, RP12-1984, Pretoria: Government Printer

changed the name of the territory to Namibia. The United Nations thus gained a direct interest in the internal affairs of Namibia, where theoretically it did not have one in the internal affairs of South Africa. This made the international campaign against apartheid easier, as apartheid was applied to what was legally an international territory.

SEPARATE DEVELOPMENT IN SOUTH WEST AFRICA

The National Party government began to apply South African apartheid laws to South West Africa after 1948. Influx controls were imposed and passes were demanded of Black migrants to the towns. Segregated suburbs were set aside for the Coloured population in addition to the already segregated Black population. Old inner Black locations, notably that in Windhoek, were demolished and rebuilt as White suburbs, while the Black population was moved several kilometres out to Katutura (Herero: 'the place where we do not settle'). However, although apartheid planning principles were introduced to the territory, the Group Areas Act was not directly applied.

In 1964 the government published the Odendaal Report on the affairs of South West Africa, and in 1968 proposed the establishment of some ten separate administrations for the various ethnically defined population groups, along the same lines as the system introduced to South Africa (Figure 7.6). The ten homelands covered some 326,000 square kilometres, or nearly 40 per cent of the area of the territory. In several cases the numbers of people involved were too small for an effective administration to be brought into being. Accordingly the representatives of the approximately 30,000 Bushmen (San) consistently sought to remain under the central government, while the fewer than 10,000 Tswanas were able to support only limited government functions.

The initial homeland was the Rehoboth Gebied which had enjoyed a measure of self-government under the Germans, followed by Ovamboland in 1967. The Ovambo population itself constituted approximately half the total of the country, and formed the numerically and politically dominant group, in marked contrast to South Africa where no Black group was so preponderant. Separate governments for Okavangoland, East Caprivi and Kaokoland followed in the early-1970s. Thereafter the administrative process continued with or without the support of elected bodies.

The development of the homeland policy, however, was entangled with the international situation. The concept of a unitary state was re-enforced as the United Nations and South Africa renewed contact, through the offices of a group of Western powers. In 1975 the government established the Constitutional Conference on South West Africa (Turnhalle Conference) based on representation by ethnic group. The homelands were thus reduced to a second tier of government, not potential independent states. In 1978 a country-wide election was held for a Constituent Assembly, which was reconstituted as the National Assembly the following year. In 1980 ethnic elections were held for the homeland administrations, except in Bushmanland, whose leaders could not support the government apparatus, the Rehoboth Gebied, where the election had taken place in 1979, and most significantly Ovamboland, where the security situation was such that

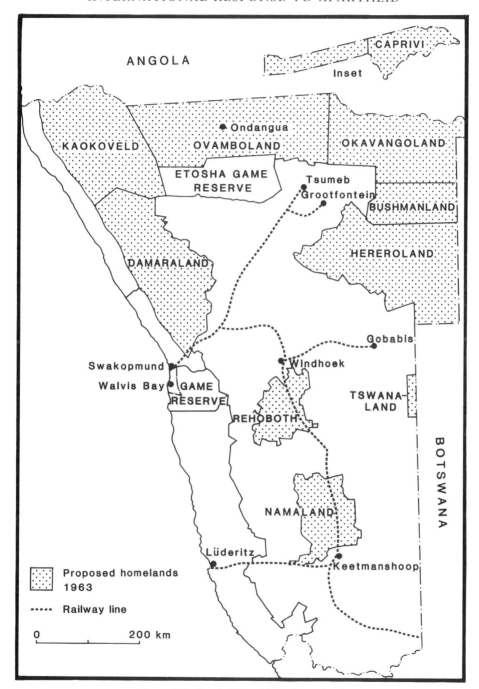

Figure 7.6 South West Africa homelands

Source After South Africa (1964) *Report of the Commission of Enquiry into South West African Affairs 1962–63*, RP12-1984, Pretoria: Government Printer

the holding of free elections was impossible.

The complex international negotiations which followed in the course of the 1980s duly led to the independent state of Namibia in 1990. During this period, although the second tier of administration was based on ethnic lines, much of the racially discriminatory legislation was repealed. Thus residential and amenity segregation was abolished in 1979. However, many social services, including education, were provided at ethnic authority level, thus retaining many aspects of apartheid until independence.

At independence there remained a number of unresolved problems between South Africa and Namibia dating from colonial days. These included the ownership of the Walvis Bay exclave, the Guano islands along the Namibian coast and the definition of the Orange River boundary.

INSURGENCY

The South African government was heavily involved in the colonial style military insurgency in South West Africa. The international complications of the insurgency were such as to lead the government into military adventures which ultimately sapped the willingness of the White electorate to continue the war. Politically the foundation of the Ovambo People's Congress in 1957 by Andimba Toivo Ya Toivo and a group of Cape Town based workers may be taken as marking the beginning of the internal challenge to continued South African rule.

The move towards the use of political violence and a more unified Black opposition began with the Old Location massacre in Windhoek in 1959, when the Ovambo People's Organisation and other parties demonstrated against forced removals to the new ethnically defined neighbourhoods outside the city. This event provided the impetus for the unification of various ethnically based organizations to form the South West African People's Organisation (SWAPO). In 1966 SWAPO began an armed insurrection against the South African administration, having organized an external base in Tanzania. As such it joined the group of diverse southern African liberation movements which were seeking the overthrow of the governments of South Africa, South West Africa, Rhodesia, Angola and Mozambique.

The South West African situation was profoundly affected by the termination of Portuguese rule and the independence of Angola in 1975. The struggle for power which ensued in the Angolan civil war was reflected in the conflict in South West Africa. Lines between the two were blurred, as SWAPO and the Angolan government fought the South African government and the UNITA movement. Indicative of the change was the removal of SWAPO's political headquarters to Lusaka and its military headquarters to Lubango (formerly Sa de Bandiera) in Angola.

In 1975 the South African army invaded Angola and occupied large sections of the south of the country, in co-operation with UNITA (Figure 7.7). However, international pressure and logistical problems resulted in withdrawal at the end of the year and a severe loss of South African military prestige.

184

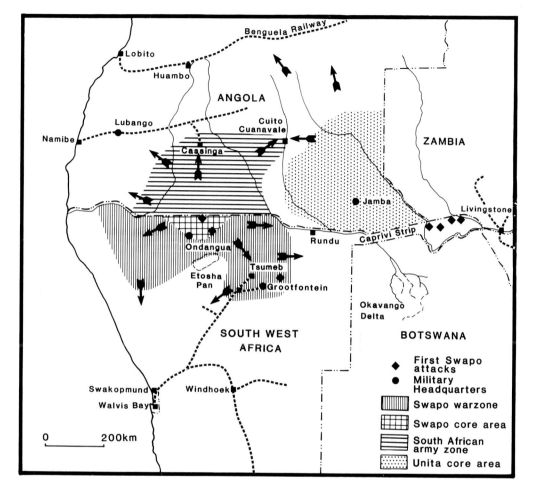

Figure 7.7 Extent of insurgency

Source Modified after Barnard, W.S. (1982) 'Die geografie van 'n revolusionere oorlog: SWAPO in Suidwes-Afrika', *South African Geographer* 10: 157–74

The militarization of South West Africa began with the raising of new forces. In contravention of the terms of the original mandate, military bases were established in the territory and local inhabitants were recruited into the South African army. The first group, the Ovambo Battalion, was followed by the formidable counter-insurgency force, Koevoet. In addition Angolans were trained as part of the South African army in a series of bases situated in the Caprivi Strip. The main aim was to keep the war firmly on the Angolan side of the border rather than allow it to develop in South West Africa itself. However, sporadic fighting took place in Ovamboland and the northern sectors of the European farming area.

DESTABILIZATION

The war in South West Africa became part of a wider struggle involving the other states in southern Africa. Moral and practical support for the ANC, PAC, and SWAPO in the neighbouring states was such that the South African government sought to exert pressure on those states to prevent their territory being used as springboards for political and military actions. These states were the particular objects of Pretoria's foreign policy after the assassination of Dr Verwoerd, through the 'outward policy' (1967–74), 'detente' (1974–8), and finally destabilization (1979–89). The 'total strategy' employed by the South African government in the 1980s involved financial and trade inducements as well as military destabilization in the hope of establishing a 'constellation of southern African states', under South African hegemony.

Destabilization involved not only intervention in the Angolan civil war at various stages between 1975 and 1989, but the effective arming of the Renamo (NMR) insurgency in Mozambique after 1980 (Figure 7.8). In both cases pressure was exerted on the new Marxist governments to revent them aiding SWAPO and the ANC. In Mozambique the South African Defence Force inherited the NMR operation from the Rhodesians who had run it in the late-1970s. Through classic insurgency tactics the NMR effectively prevented the Mozambiquan government from gaining full control over its territory, and contributed to the national bankruptcy brought about by the adoption of Communist principles. Large-scale refugee movements led to further disruptions. In Angola the ability of the South African army to prevent the defeat of UNITA and the widespread destruction of the civil war had similar results.

In 1984 the South African government entered into agreements with the Angolan government in Lusaka, and the Mozambiquan government through the Nkomati Accord. The agreements provided for the end of destabilization and the withdrawal of South African troops from Angola in return for assurances that the two countries would not be used for attacks on South Africa or South West Africa. The diplomatic success was shortlived as South African destabilization activities continued as did SWAPO and ANC actions. Thus the completion of the South African withdrawal from Angola in 1985 coincided with the capture of a South African commando unit involved in a sabotage mission against the Cabinda oil wells, in northern Angola.

Destabilization was also directed towards other southern African states and towards the sabotaging of negotiations which were deemed unfavourable to the South African government. Lesotho was particularly susceptible to pressures. A major raid on Maseru in 1982 was aimed specifically at destroying the ANC presence in the country. In 1986 the Lesotho government was overthrown as a result of pressures exerted by South Africa through a slowdown in border control formalities, resulting in the paralysis of trade and population movements over the common border. A more amenable regime was duly installed. Raids on targets in Gaborone, Harare and Lusaka in 1986 effectively ended the Commonwealth Eminent Persons initiative, seeking a solution to South Africa's internal problems. In 1987 raids on Gaborone and Livingstone indicated

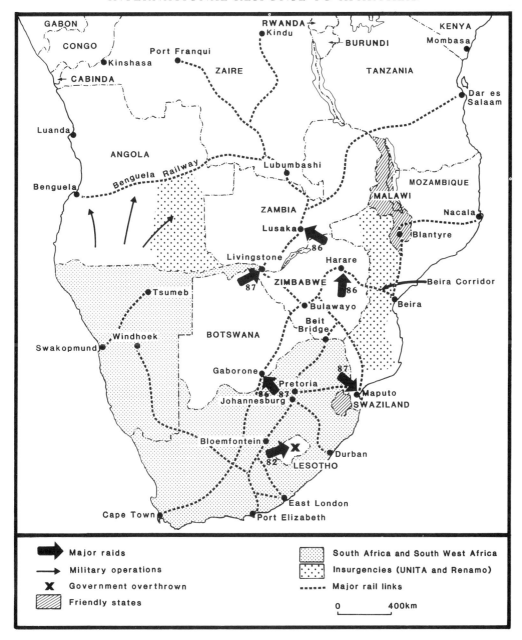

Figure 7.8 Elements of destabilization

Source Compiled from various newspaper sources

that ANC personnel would be targeted anywhere in the subcontinent. Militarily at least, South African destabilization succeeded in preventing overt action by its neighbours and keeping the ANC and SWAPO bases at some distance from South Africa's borders. However, the heavy financial and political costs of the policy were such that the government reverted to peace initiatives in 1988.

THE SOUTH AFRICAN DEFENCE FORCE

Destabilization was made possible by the numerical and technical dominance of the South African Defence Force in the subcontinent. After 1948 the permanent military establishment had initially been viewed with suspicion by the National Party, as many Afrikaner nationalists had actively opposed the South African entry into the Second World War, and some had supported the cause of National Socialist Germany. In addition the Party was opposed to the arming of Blacks and Coloureds, believing that this would weaken the position of the Whites, who alone were to be allowed to bear arms.

However, the perceived threat from countries elsewhere in Africa and the imposition of an arms embargo by Great Britain and the United States in 1963 led to a reassessment of the situation. The deterioration of the security situation in South West Africa necessitated an enlargement of the army, to augment the police force. The military call-up for White men was extended to two years by 1977 with substantial periods of duty in the reserves thereafter. The length of these duties was such that 7 per cent of all White males aged between 18 and 45 were engaged in the military service by 1980. This was not a high proportion if compared with Israel or a number of militarized states elsewhere in the world, but it was high for western democracies, with which the White population usually identified. This was to be one of the major causes of dissatisfaction with the National Party amongst the White population by the late-1980s.

In a major departure in principle the government decided to arm the Coloured volunteers to the Cape Corps in 1973. In addition Black units were raised for the Transkei in 1974, to be followed by other homeland units thereafter. Black units in the South African Defence Force were initiated in 1974 in South West Africa and in South Africa soon afterwards. By 1980 approximately half the army was no longer White. Ten years later the proportion had risen to over three-quarters.

In terms of size the South African Defence Force was larger, and better trained, than any other in the region (Figure 7.9). It was capable of being augmented with reserves in times of crisis to provide overwhelming numerical superiority. Consequently by the late-1970s large numbers of men could be deployed for long periods of time in South West Africa or in the Black townships of South Africa. It is worth noting that by 1980 SWAPO probably had no more than 2,000 men in the field at any given time, yet it required 20,000 to 25,000 members of the South African Defence Force to counter them.

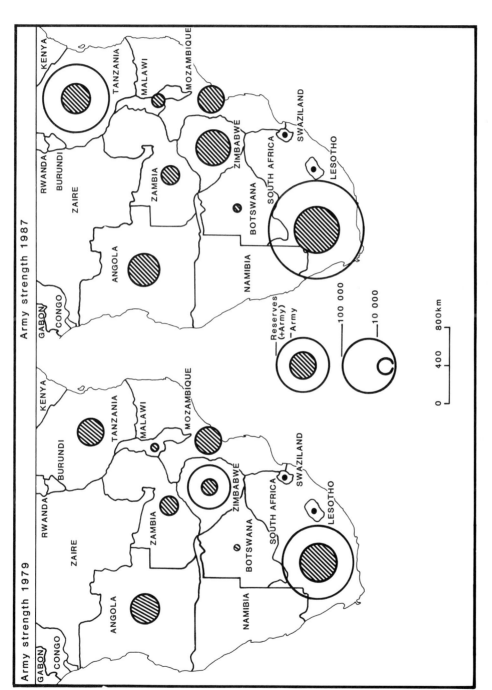

Figure 7.9 Regional army strengths, 1979 and 1987

Source Based on figures from the Statesman's Yearbook for 1979–87

CONSTELLATION OF SOUTHERN AFRICAN STATES

The South African government pursued destabilization and open warfare as part of a policy to establish security. An equally significant aspect was the concept of the constellation of southern African states bound together economically, which pursued politically different, but not antagonistic courses. In part this was related to the thinking that the way to international acceptance was through African co-operation.

The concept of a southern African commonwealth was first suggested by Dr H.F. Verwoerd in 1963. It was his successor, Mr B.J. Vorster, who actively pursued an outward policy, aimed at building a ring of friendly states between South Africa and the remainder of the continent. South Africa, Rhodesia and the Portuguese colonies drew closer together both economically and diplomatically as outside hostility increased. Practical co-operation included such major schemes as the Cahora Bassa hydro-electric scheme and the Ruacana Water scheme (Figure 7.10). In both cases the Portuguese colonies were partially integrated into the South African economy and security system. However, one of the major casualties of the subsequent destabilization era was the destruction of the power lines from Cahora Bassa to the eastern Transvaal.

Only one independent state in southern Africa co-operated fully with the South African government in its outward policy, namely Malawi. President Banda sought aid from wherever he could. South African finance thus became available for a variety of projects, of which the building of the new capital city at Lilongwe was the most spectacular.

The independence of Angola and Mozambique in 1975, as in much else, caused major shifts in South African foreign policies, as the cordon sanitaire was effectively breached and hostile states shared common borders with the country. The crisis in Rhodesia became more pressing as the Mozambiquan government closed the border between the two states and Rhodesian trade was redirected through Botswana. An emergency railway link was subsequently built through Beit Bridge to connect the two countries directly.

The concept of a constellation of southern African states was revived by Mr P.W. Botha in 1977. In the following year the Development Bank of Southern Africa was established to offer South African aid in funding suitable projects in the subcontinent. Nevertheless, the Rhodesian war prevented any meaningful *detente* in southern Africa. The ending of South African support for the UDI government in Rhodesia cleared the way for the establishment of the grouping. However, the subsequent success of the ZANU–PF in winning the first elections for an independent Zimbabwe effectively ended the concept of a South African-led southern African grouping.

Following the independence of Zimbabwe in 1980 the various frontline states in southern Africa formed the Southern African Development Coordination Conference (SADCC) at a meeting in Lusaka. The express purpose of the organization was the lessening of the states' economic dependence upon South Africa. The nine countries, increased to ten with the independence of Namibia ten years later, in particular sought

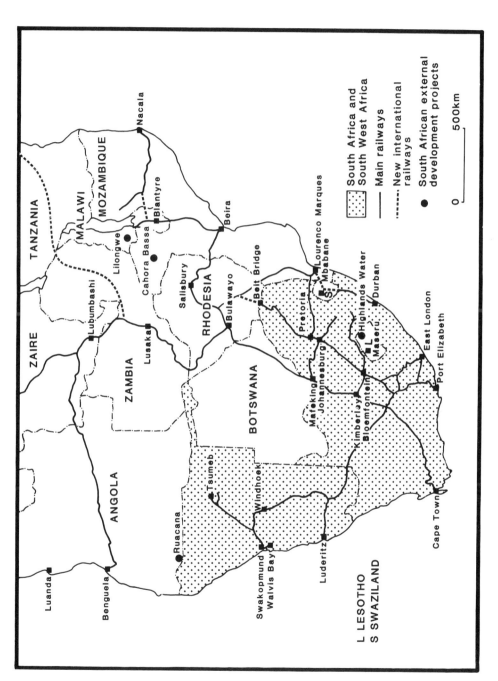

Figure 7.10 Constellation of Southern African States

Source Compiled from various newspaper sources

South Africa and
South West Africa

Main railways

New international
railways

● South African external
development projects

0   500km

L LESOTHO
S SWAZILAND

to decrease their dependence on the South African transportation system. In 1980 South Africa accounted for 76 per cent of the Gross National Product of the region, and was a major trading partner of all but Tanzania and Angola (Figure 7.11).

The particular problems of Botswana, Lesotho and Swaziland as members of the South African Customs Union are evident. The other countries exhibited greater independence, although in 1980 the Zimbabwean economy was heavily dependent upon the South African economy as a result of the international trade sanctions imposed on Rhodesia after the unilateral declaration of independence in 1965.

MINE LABOUR

The South African mining industry developed a wide recruiting zone in southern Africa based on the principle of employing single men as temporary migrants. After the Anglo-Boer War at the turn of the century it was the recruitment of labourers from Mozambique which effectively supplied the manpower to re-establish the mines and

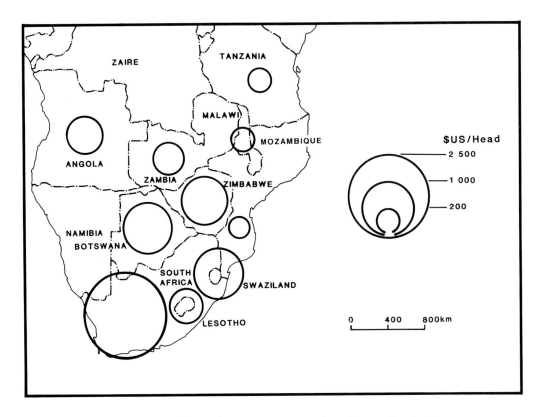

Figure 7.11 Southern Africa: Gross National Product per head, 1984

Source Based on figures from the Statesman's Yearbook for 1984

boost production. Health problems with those recruited from central Africa led to restrictions on employing men from north of 22°S until 1932, while the governments of Southern and Northern Rhodesia attempted to retain as much labour for their own economies as possible, and thus discouraged workers from going to South Africa. The scale of the demand was such that the local supply of labour remained inadequate until the 1970s and an intensive search to gain men from other parts of the subcontinent ensued. After the 1970s the employers' need to diversify the sources of supply, to reduce reliance on any one source, resulted in the continued employment of foreign labour.

The National Party policies initially had little impact upon mine labour recruitment (Figure 7.12). Thus the patterns of supply in 1960 essentially reflected those of 1948. Changes came with African independence, as the more northerly countries halted recruitment to demonstrate their opposition to the migratory labour system and the South African government. However, the mining houses' demands for more labour in the prosperous 1960s resulted in a boost in numbers and the development of Malawi as the main source of foreign manpower for the South African mines. In 1974 an air crash in Botswana resulted in the death of seventy-five Malawian migrant workers and recruitment was suspended. The civil war in Mozambique resulted in a severe curtailment of the supply from that country. As a result the number of foreign miners in South Africa fell from an average of 316,000 in 1973 to 196,000 in 1977, at which level it remained fairly constant thereafter.

In 1978 the Southern African Labour Commission was established by the countries supplying labour to South Africa with the aim of eliminating the migratory system. However, little action was taken. Only Zimbabwe completely disengaged after independence, thereby fulfilling the aims of earlier colonial governments. The system of deferred payments, whereby part, sometimes as much as 60 per cent, of the migrants' earnings were remitted to the government concerned, gave the supplying governments a vested interest in the system's continuance. Thus, far from being able to exert pressure on South Africa as envisaged, the South African recruitment agencies were able to exert pressure on the source countries.

The reduction in the number of foreign workers was made up with local labour and men recruited from Lesotho, which by 1976 had become the largest supplier of labour to South Africa's mines. Mine labourers thus migrated over shorter distances than before, enabling many to return home at weekends, placing them in the same position as homeland migrants.

STRATEGIC INDUSTRIES

Economic sanctions proved to be the most effective weapon of the international community against apartheid. A voluntary arms embargo was imposed by the United Nations in 1963. Great Britain and the United States ceased their trade, but their place was taken by France and other countries. In 1977 the arms embargo became mandatory,

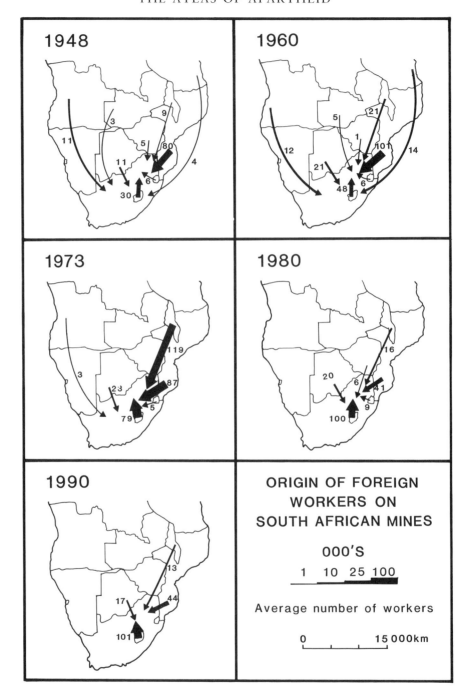

Figure 7.12 Sources of mine labour, 1948–90

Source Compiled from statistics in the annual reports of the Chamber of Mines

but by that time the country had become largely self-sufficient in production of those weapons required for the types of war waged by the South African Defence Force.

Strategic projects, including the development of SASOL (South African Coal, Oil and Gas Corporation) were initiated after 1948. SASOL was created at what became Sasolburg and was enhanced with the construction of Secunda in the 1970s and 1980s for the extraction of oil from coal. The aim was to establish strategic self-sufficiency in oil production. The Mossel Bay oil and gas project in the 1980s and early-1990s similarly sought to lessen South Africa's dependence upon imported oil. However, the cost of the latter exceeded R10 billion with little chance of the project ever showing a profit.

Other such projects included the investment in local content for motor vehicles, more especially the diesel engine plant at Atlantis, together with the armament industry, including the manufacture of aircraft and coastal vessels (Figure 7.13). Armscor

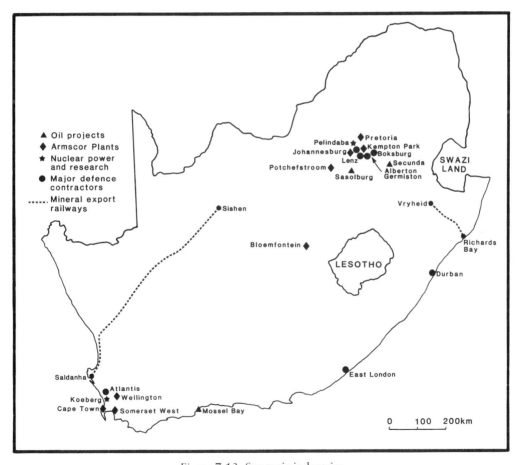

Figure 7.13 Strategic industries

Source Various sources, including Rogerson, C.M. (1990) 'Defending apartheid: Armscor and the geography of military production in South Africa', *GeoJournal* 22: 241–50

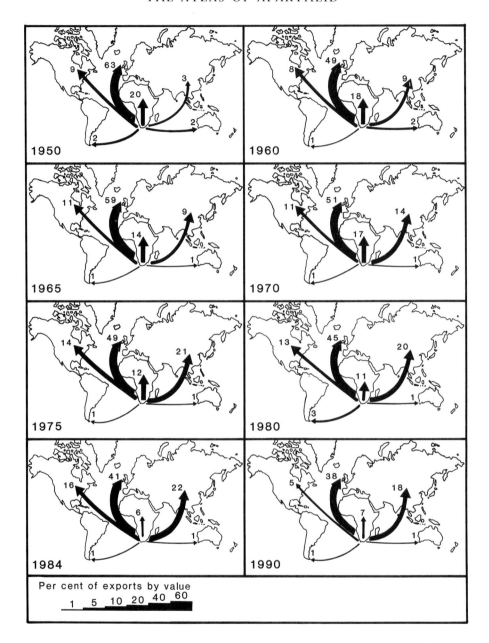

Figure 7.14 Exports from South Africa, 1950–90, by continent

Source Compiled from the Statistical Yearbook of South Africa (various dates) and statistics released by the Department of Trade and Industry

(originally the Armaments Production Board) was formed in 1964 to counter the United Nations arms embargo with a programme of strategic industrial development. The corporation was involved, through a number of subsidiaries, in the production of a wide range of military equipment ranging from aircraft production at Kempton Park, to explosives and projectiles in the western Cape. In addition Armscor contracted out a wide range of products to private firms. The main areas of activity manufacture were located in the Witwatersrand and Pretoria. Government subsidies for these schemes were such that no economic return could be expected and ultimately they helped to undermine the South African economy.

Mineral export promotion projects were also developed. Coal exports from the eastern Transvaal and iron ore from the northern Cape were designed to add to South Africa's foreign earning capacity. Special railways from the mines to the newly constructed ports of Saldanha and Richards Bay were among the major schemes undertaken in the 1970s.

TRADE SANCTIONS

In other spheres trade sanctions have been irregular in imposition and effect. India first imposed sanctions in 1946 in response to the anti-Indian legislation of the same year. Further countries banned specific forms of economic activity, such as direct investment by Japan in 1964, which did not preclude the expansion of trade between the two states. Sweden, a leading political opponent of apartheid, only forbade new investment in 1979.

It was in response to the riots of 1984 that comprehensive measures were widely adopted against South Africa. The United Nations Security Council, the European Community, the Commonwealth and the United States all imposed voluntary embargoes within the next few years. Probably the most influential was the enactment of the American Comprehensive Anti-Apartheid Act of 1986. Credit squeezes, oil embargoes and the restriction of precious and strategic metal exports all followed. Disinvestment became serious and trade, particularly American trade figures, fell markedly. The South African economy was thus placed under substantial pressure reducing economic growth rates in the late-1980s to near zero, while the population continued to expand at approximately 2.5–3.0 per cent per annum. The countries reducing their imports from South Africa were led by the United States, whose trade halved between the early-1980s and 1990. Other countries, notably Germany, Japan and Taiwan made up a great deal of the shortfall (Figure 7.14). External pressures thus contributed significantly to the final demise of apartheid.

FURTHER READING

The literature on apartheid is extensive, ranging from the highly philosophical questioning of the origins of the policy, and political analyses of its implementation, to specific case studies of particular communities or pieces of land. What follows is more explicitly concerned with its spatial manifestations. The publications of the South African Institute of Race Relations (Johannesburg), and more particularly its annual *Survey of Race Relations* published since 1951, provides an immense fund of information, as do those of the Africa Institute (Pretoria) and the Development Bank of Southern Africa (Midrand). The various government reports, together with answers to questions in parliament, published in the parliamentary *Hansard* (Cape Town) also provide inestimable factual information. Regrettably the coverage is patchy and in some areas non-existent, reflecting the racially selective nature of much of the writing and data collection in the country.

GENERAL

Adam, H. and Giliomee, H. (1979) *Ethnic Power Mobilized: Can South Africa Change?*, New Haven: Yale University Press.

Christopher, A.J. (1982) *South Africa*, London: Longman.

Davenport, T.R.H. (1991) *South Africa: A Modern History*, London: Macmillan.

Drakakis-Smith, D. (ed.) (1992) *Urban and Regional Change in Southern Africa*, London: Routledge.

Horowitz, D.L. (1991) *A Democratic South Africa? Constitutional Engineering in a Divided Society*, Berkeley: University of California Press.

Inskeep, R.R. (1978) *The Peopling of Southern Africa*, Cape Town: David Philip.

Lemon, A. (1976) *Apartheid: A geography of Separation*, Farnborough: Saxon House.

—— (1987) *Apartheid in Transition*, Aldershot: Gower.

Saunders, C. (ed.) (1988) *Illustrated History of South Africa: The Real Story*, Cape Town: Reader's Digest.

Smith, D.M. (ed.) (1982) *Living Under Apartheid*, London: George Allen & Unwin.

South African Review (1983–92) Johannesburg: Raven Press.

Thomas, F.A. (ed.) (1981) *South Africa: Time Running Out*, Berkeley: University of California Press.

FURTHER READING

BEFORE APARTHEID

Bundy, C. (1979) *The Rise and Fall of the South African Peasantry*, London: Heinemann.

Christopher, A.J. (1976) *Southern Africa: Studies in Historical Geography*, Folkestone: Dawson.

—— (1988) *The British Empire at its Zenith*, London: Croom Helm.

Elphick, R. and Giliomee, H. (1979) *The Shaping of South African Society 1652–1820*, Cape Town: Longman.

Lamar, H. and Thompson, L. (eds) (1981) *The Frontier in History: North America and Southern Africa Compared*, New Haven: Yale University Press.

Letsoalo, E.M. (1987) *Land Reform in South Africa*, Johannesburg: Skotaville Publishers.

Mabin, A. (1986) 'Labour, capital, class struggle and the origins of residential segregation in Kimberley 1880–1920', *Journal of Historical Geography*, 12:4–26.

Pelzer, A.N. (1966) *Verwoerd Speaks: Speeches 1948–1966*, Johannesburg: APB Publishers.

ADMINISTERING APARTHEID

Best, A.C.G. and Young, B.S. (1972) 'Capitals of the homelands', *Journal for Geography*, 3:1043–55.

Browett, J.G. (1976) 'The application of a spatial model to South Africa's development regions', *South African Geographical Journal*, 58:118–29.

Fair, T.J.D. (1982) *South Africa: Spatial Frameworks for Development*, Cape Town: Juta.

Heymans, C. and Totemeyer, G. (eds) (1988) *Government by the People: The Politics of Local Government in South Africa*, Cape Town: Juta.

Seethal, C. (1991) 'Restructuring the local state in South Africa: Regional Services Councils and crisis resolution', *Political Geography Quarterly*, 10:8–25.

STATE APARTHEID

Bell, T. (1973) *Industrial Decentralisation in South Africa*, Cape Town: Oxford University Press.

Blenck, J. and von der Ropp, K. (1977) 'Republic of South Africa: is partition a solution?' *South African Journal of African Affairs*, 7:21–32.

Christopher, A.J. (1982) 'Partition and population in South Africa', *Geographical Review*, 72:127–38.

du Toit, B.M. (1991) 'The far right in current South African politics,' *Journal of Modern African Studies*, 29:627–67.

Esterhuysen, P. (1982) 'Greater Swaziland?' *Africa Insight*, 12:181–8.

Griffiths, I. and Funnell, D.C. (1991) 'The abortive Swaziland deal,' *African Affairs*, 90:51–64.

Malan, T. And Hattingh, P.S. (eds) (1976) *Black Homelands in South Africa*, Pretoria: Africa Institute.

Moolman, H.J. (1977) 'The creation of living space and homeland consolidation with reference to Bophuthatswana', *South African Journal of African Affairs*, 7:149–63.

Murray, C. and O'Regan, C. (eds) (1990) *No Place to Rest: Forced Removals and the Law in South Africa*, Cape Town: Oxford University Press.

Platzky, L. and Walker, C. (1985) *The Surplus People: Forced Removals in South Africa*, Johannesburg: Ravan.

FURTHER READING

Rotberg, R.I. and Barratt, J. (eds) (1980) *Conflict and Compromise in South Africa*, Cape Town: David Philip.

South Africa (1955) *Summary of the Report of the Commission for the Socio-Economic development of the Bantu Areas within the Union of South Africa*, UG61–1955 (Tomlinson Report), Pretoria: Government Printer.

—— (1960) *Report of the Commission of Inquiry into European Occupancy of the Rural Areas*, Pretoria: Government Printer.

Stern, E. (1987) 'Competition and location in the Gaming Industry: the "Casino States" of Southern Africa,' *Geography*, 72:140–50.

Van Onselen, C. (1990) 'Race and class in the South African countryside: cultural osmosis and social relations in the share cropping economy of the South-Western Transvaal 1900–1950', *American Historical Review*, 95:99–123.

URBAN APARTHEID

Christopher, A.J. (1989) 'Apartheid within apartheid: an assessment of official intra-Black segregation on the Witwatersrand, South Africa', *Professional Geographer*, 41:328–36.

—— (1990) 'Apartheid and urban segregation levels in South Africa', *Urban Studies*, 27:421–40.

—— (1991) 'Changing patterns of group area proclamations in South Africa 1950–1989', *Political Geography Quarterly*, 10:240–53.

Davies, R.J. (1981) 'The spatial formation of the South African city', *GeoJournal*, Supplementary Issue, 2:59–72.

de Coning, C., Fick, J. and Olivier, N. (1986) *Residential Settlement Patterns: a Pilot Study of Socio-Political Perceptions in Grey Areas of Johannesburg*, Johannesburg: Rand Afrikaans University.

Krige, D.S. (1988) *Die Transformasie van die Suid-Afrikaanse Stad*, Bloemfontein: University of the Orange Free State.

Kuper, L., Watts, H. and Davies, R. (1958) *Durban: A Study in Racial Ecology*, London: Jonathan Cape.

Lemon, A. (ed.) (1991) *Homes Apart: South Africa's Segregated Cities*, London: Paul Chapman.

Mather, C. (1987) 'Residential segregation and Johannesburg's "Locations in the sky"', *South African Geographical Journal*, 69:119–28.

Morris, P. (1980) *Soweto: A Review of Existing Conditions and Some Guidelines for Change*, Johannesburg: Urban Foundation.

Nel, J. (1988) *Die Geografiese Impak van die Wet op Groepsgebiede en Verwante Wetgewing op Port Elizabeth*, Port Elizabeth: University of Port Elizabeth.

Rule, S.P. (1989) 'The emergence of a racially mixed residential suburb in Johannesburg: demise of the apartheid city?' *Geographical Journal*, 155:196–203.

Smith, D.M. (ed.) (1992) *The Apartheid City and Beyond: Urbanization and Social Change in South Africa*, London: Routledge.

PERSONAL APARTHEID

Mattera, D. (1987) *Memory is the Weapon*, Johannesburg: Ravan.

FURTHER READING

Pinnock, D. (1984) *The Brotherhoods: Street Gangs and State Control in Cape Town*, Cape Town: David Philip.

Western, J. (1981) *Outcast Cape Town*, Minneapolis: University of Minnesota Press.

RESISTANCE TO APARTHEID

Cobbett, W. and Cohen, K. (1988) *Popular Struggles in South Africa*, London: James Currey.

Davis, S.M. (1987) *Apartheid's Rebels: Inside South Africa's Hidden War*, New Haven: Yale University Press.

Meli, F. (1988) *South Africa Belongs to Us: a History of the ANC*, Harare: Zimbabwe Publishing House.

South Africa (1980) *Report of the Inquiry into the riots at Soweto and Elsewhere*, RP55–1980, Pretoria: Government Printer.

INTERNATIONAL RESPONSE TO APARTHEID

Berat, L. (1990) *Walvis Bay: Decolonization and International Law*, New Haven: Yale University Press.

Dixon, C. and Heffernan, M.J. (eds) (1991) *Colonialism and Development in the Contemporary World*, London: Mansell.

Geldenhuys, D. (1990) *Isolated States: A Comparative Analysis*, Johannesburg: Jonathan Ball.

Gibb, R.A. (1987) 'The effect on the countries of SADCC of economic sanctions against the Republic of South Africa', *Transactions of the Institute of British Geographers*, 12:398–412.

Griffiths, I. (1989) 'Airways sanctions against South Africa', *Area*, 21:249–59.

Herbstein, D. and Evenson, J. (1989) *The Devils are Among Us: The War for Namibia*, London: Zed Books.

Hyam, R. (1972) *The Failure of South African Expansion*, London: Macmillan.

Pirie, G.H. (1990) 'Aviation, apartheid and sanctions: air transport to and from South Africa, 1945–1989', *GeoJournal*, 22:231–40.

Ramphal, S. (1986) *Mission to South Africa: The Commonwealth Report*, Harmondsworth: Penguin.

—— (1989) *South Africa: The Sanctions Report*, Harmondsworth: Penguin.

Rogerson, C.M. (1990) 'Defending apartheid: Armscor and the geography of military production in South Africa', *GeoJournal*, 22:241–50.

South Africa (1964) *Report of the Commission of Enquiry into South West African Affairs 1962–63*, RP12–1964, Pretoria: Government Printer.

NOTES

INTRODUCTION

1 South Africa, *Hansard 1954*, col 5854, 31 May 1954.
2 South Africa, *Hansard 1950*, col 7722, 31 May 1950.
3 South Africa, *Hansard 1949*, col 6164, 19 May 1949.
4 South Africa, *Hansard 1953*, col 1053, 6 August 1953.
5 South Africa, *Hansard 1953*, cols 1054–55, 6 August 1953.
6 South Africa, *Hansard 1953*, col 3576, 17 September 1953.
7 W.L. Shirer, *The Rise and Fall of the Third Reich* (London: Pan, 1964), p. 1118.
8 D. Posel, *The Making of Apartheid 1948–1961: Conflict and Compromise* (Oxford: Clarendon Press, 1991).
9 *Financial Mail* (Johannesburg), 19 August 1977.
10 South Africa, *Hansard 1959*, col 6221, 20 May 1959.
11 *Eastern Province Herald* (Port Elizabeth), 7 December 1977.
12 Television speech 18 March 1992.
13 *Weekly Mail* (Johannesburg), 29 May–4 June 1992.

1 BEFORE APARTHEID

1 Transvaal, *Report of the Transvaal Local Government Commission, T.P. 1–1922* (Pretoria: Government Printer, 1922), p. 42.
2 Figure 1.23. The coverage presented may be incomplete but is designed to present as comprehensive a picture as possible. In 1911 many locations were informal arrangements with no governmental backing. However, they were *de facto* entities so far as census enumerators were concerned and this source is likely to be more comprehensive than the *de jure* position.
3 Durban, *Mayor's Minute* (Durban: Durban Corporation, 1923), p. 21.
4 See Deeds Office, Pietermaritzburg, Deed of Transfer 27/1921, Durban.
5 See Deeds Office, Cape Town, Deed of Transfer 9672/1919, Fish Hoek.
6 In Figures 1.27–1.30, indices of segregation are calculated according to the following formula:

$$I_x = \frac{0.5\Sigma \mid x_i - z_i \mid}{1 - \dfrac{X}{Z}}$$

where: l_x is the Index of Segregation of population X, X represents the total of subgroup X in the city, Z represents the total population of the city, x_i represents the percentage of the X population in the ith area, and Z_i represents the percentage of the total population in the ith area. The resultant index is expressed on a scale from zero (complete integration) to 100 (complete segregation).

3 STATE APARTHEID

1 South Africa *Summary of the Report of the Commission for the Socio-Economic Development of the Bantu Areas within the Union of South Africa* (Pretoria: Government Printer, 1955).

4 URBAN APARTHEID

1 South Africa, *Hansard 1950*, col. 7722, 31 May 1950.
2 W.R.P. de Swardt 'Urbanism and Bantu – focus on town planning and development in Soweto', pp. 467–76 in H.L. Watts (ed.) *Focus on Cities* (Durban: University of Natal, Institute of Social Research, 1970).

5 PERSONAL APARTHEID

1 South Africa, *Hansard 1953*, col 3576, 17 September 1953.
2 South Africa, *Hansard 1959*, col 6221, 20 May 1959.
3 *Weekly Mail* (Johannesburg), 30 April 1992.

6 RESISTANCE TO APARTHEID

1 *Time* (New York), 26 September 1977.

7 INTERNATIONAL RESPONSE TO APARTHEID

1 South Africa *Report of the Commission of Enquiry into South West African Affairs 1962–63*, (Pretoria: Government Printer, 1964), p. 47.

INDEX

INDEX